Implementing Identity Management on GCP

Learn to Solve Customer and Workforce IAM Challenges on GCP

Advait Patel
Saai Krishnan Udayakumar
Hariharan Ragothaman

apress®

Implementing Identity Management on GCP: Learn to Solve Customer and Workforce IAM Challenges on GCP

Advait Patel
Software Engineering, Broadcom,
Glenview, IL, USA

Saai Krishnan Udayakumar
Washington, WA, USA

Hariharan Ragothaman
Massachusetts, MA, USA

ISBN-13 (pbk): 979-8-8688-1696-3
https://doi.org/10.1007/979-8-8688-1697-0

ISBN-13 (electronic): 979-8-8688-1697-0

Copyright © 2025 by Advait Patel, Saai Krishnan Udayakumar and Hariharan Ragothaman

This work is subject to copyright. All rights are reserved by the Publisher, whether the whole or part of the material is concerned, specifically the rights of translation, reprinting, reuse of illustrations, recitation, broadcasting, reproduction on microfilms or in any other physical way, and transmission or information storage and retrieval, electronic adaptation, computer software, or by similar or dissimilar methodology now known or hereafter developed.

Trademarked names, logos, and images may appear in this book. Rather than use a trademark symbol with every occurrence of a trademarked name, logo, or image we use the names, logos, and images only in an editorial fashion and to the benefit of the trademark owner, with no intention of infringement of the trademark.

The use in this publication of trade names, trademarks, service marks, and similar terms, even if they are not identified as such, is not to be taken as an expression of opinion as to whether or not they are subject to proprietary rights.

While the advice and information in this book are believed to be true and accurate at the date of publication, neither the authors nor the editors nor the publisher can accept any legal responsibility for any errors or omissions that may be made. The publisher makes no warranty, express or implied, with respect to the material contained herein.

Managing Director, Apress Media LLC: Welmoed Spahr
Acquisitions Editor: Aditee Mirashi
Coordinating Editor: Jacob Shmulewitz

Distributed to the book trade worldwide by Springer Science+Business Media New York, 1 New York Plaza, New York, NY 10004. Phone 1-800-SPRINGER, fax (201) 348-4705, e-mail orders-ny@springer-sbm.com, or visit www.springeronline.com. Apress Media, LLC is a Delaware LLC and the sole member (owner) is Springer Science + Business Media Finance Inc (SSBM Finance Inc). SSBM Finance Inc is a **Delaware** corporation.

For information on translations, please e-mail booktranslations@springernature.com; for reprint, paperback, or audio rights, please e-mail bookpermissions@springernature.com.

Apress titles may be purchased in bulk for academic, corporate, or promotional use. eBook versions and licenses are also available for most titles. For more information, reference our Print and eBook Bulk Sales web page at http://www.apress.com/bulk-sales.

Any source code or other supplementary material referenced by the author in this book is available to readers on GitHub (https://github.com/Apress). For more detailed information, please visit https://www.apress.com/gp/services/source-code.

If disposing of this product, please recycle the paper

*To my parents, **Nikunj and Rajesh Patel**, for your unwavering love and belief that shaped the person I am today; to my wife, **Charmi Patel**, for your constant strength, support, and faith in every step of my journey; to my uncle, **Prashant Patel**, whose guidance and steadfast presence have been a quiet pillar of strength throughout my life; and to my son, **Avyakt Patel**, may this work be a part of the legacy I build for you, with all the love and hope a father holds for the future.*

—Advait Patel

*To my son **Darsh Krishnan** whose smile keeps me going; and to my wife **Ramya Sudarsan** for being my pillar through every high and low.*

—Saai Krishnan Udayakumar

*To my mother, **Jayalakshmi Ragothaman**, whose love continues to guide me; to my wife, **Aishwarya Murali**, my steadfast anchor; and to **Beacon**, whose joy brightens even the darkest days.*

—Hariharan Ragothaman

Table of Contents

About the Authors xi

About the Technical Reviewer xiii

Acknowledgments xv

Introduction xvii

Chapter 1: Introduction to Identity and Access Management on Google Cloud Platform 1

   Overview of Identity and Access Management 1

      Benefits of IAM 3

      IAM in Modern Organizational Environments 4

      Future Directions of IAM 4

   Overview of IAM on GCP 5

      Identities 5

      Resources 6

      Roles 7

      Policies 7

   Real-World Significance of IAM 8

      IAM Use in Fintech: Fostering Security and Compliance 8

      Use of IAM in Healthcare to Protect Patient Data and Enable Continued Access 9

      IAM in Gaming to Enhance User Experience and Address Cheating 10

   Tools Required 11

   Emerging Trends 12

   References 14

Chapter 2: Managing Roles, Policies, and Permissions 17

Types of Roles 17
Predefined Roles 17
Custom Roles 19
Basic Roles 20
Best Practices in Role Management in GCP IAM 21

IAM Policies and Inheritance in the GCP Resource Hierarchy 21
IAM Policies in GCP 22
GCP Resource Hierarchy 22
IAM policy inheritance 24
Policy Evaluation in the Resource Hierarchy 24
Overriding IAM Policies 25
Best Practices for Inheritance and IAM Policies 25

Real-World Use Case: Managing Permissions in a Multiproject Organization 26
Spotify Challenge 26
Solution 27
Spotify's Outcome 28

Real-World Insights: Lessons Learned from IAM Misconfigurations in Large Enterprises 28

References 30

Chapter 3: Advanced IAM Features 33

Using IAM Conditions for Context-Aware Access 33
IAM Conditions 33
IAM Conditions and Context-Aware Access 34
Benefits of Using IAM Conditions 36
Best Practices in IAM Conditions 37

Cross-Project and Cross-Organization Access with IAM Policies 38
Cross-Project and Cross-Organization Access 38
Enabling Cross-Project and Cross-Organization Access 40
Benefits of Enabling Cross-Project and Cross-Organization Access 41
Best Practices for Cross-Project and Cross-Organization Access 42

Implementing Temporary and Time-Bound Access Securely 43
    Demand for Temporary and Time-Bound Access 44
    Strategies for Secure Time-Bound Access 44
Industry-Specific Examples of Context-Aware Access Policies in Healthcare and Education 46
    Context-Aware Access in Healthcare 46
    Context-Aware Access in Education 47
References 48

Chapter 4: Service Accounts and Workload Identity 51

I. Service Accounts: Types and Best Practices 51
    Types of Service Accounts 51
    Best Practices in Service Accounts 55
II. Workload Identity Federation for Secure Service Access 57
    Benefits of Workload Identity Federation 59
    Enabling WIF 59
    Real-World Use 60
    Implementing Workload Identity Federation 61
III. Managing and Rotating Service Account Keys 62
    Service Account Key Rotation 63
    Strategies for Key Management 65
References 67

Chapter 5: Securing API and Workloads 69

I. IAM Permissions for API Gateway, Cloud Functions, and Cloud Run 69
    IAM Use in Google Cloud 69
    IAM Permissions in Using API Gateway 70
    IAM Permissions for Cloud Functions 71
    IAM Permissions for Cloud Run 72
    Managing Service Accounts 73
    Troubleshooting IAM Issues 74
II. Service-to-service Authentication Using OAuth2.0 and API Keys 74
    Service-to-Service Authentication 75

TABLE OF CONTENTS

    Using OAuth 2.0 for Service-to-Service Authentication 75

    API Keys for Service-to-Service Authentication 77

  III. Real-World Example: Securing an API-Based SaaS Product 80

  References 81

Chapter 6: Automating IAM Policies 83

  I. Using Terraform to Manage IAM at Scale 83

    Setting Up Terraform for GCP IAM Management 84

  II. Automating Workflows with Scripts and the IAM API 89

  III. Real-World Governance Strategies for Large Enterprises 93

  IV. Pro Tips for Automating Policy Updates Efficiently 96

  References 96

Chapter 7: Auditing and Monitoring IAM Policies 99

  I. Using Cloud Audit Logs to Track IAM Policy Changes 99

    Understanding Cloud Audit Logs 99

    Enabling Cloud Audit Logs 100

    Tracking IAM Policy Changes with Cloud Audit Logs 100

    Best Practices in Using Cloud Audit Logs to Track IAM Policy Changes 102

  II. Setting Up IAM Alerts with Cloud Monitoring 103

    Cloud Monitoring for IAM 105

    Requirements for Setting Up IAM Alerts 105

    Process of Setting Up IAM Alerts in Cloud Monitoring 105

    Advanced Alerting Strategies 106

    Best Practices to Be Used in IAM Alerting Activities 107

  III. Leveraging Cloud Asset Inventory for Compliance 108

    Best Practices to Assert Compliance with CAI 111

  IV. Case Study: How a Healthcare Organization Uses IAM Monitoring to Meet HIPAA Compliance 112

  References 113

Chapter 8: Managing Multicloud and Hybrid IAM 115

I. Integrating GCP IAM with AWS and Azure for Hybrid Environments 115

    Integration Strategies 116

    Best Practices in Securing Multicloud IAM Integration 118

II. Best Practices for Cross-Cloud IAM Governance 120

    Implementation of Centralized Identity Provider (IdP) 120

    Principle of Least Privilege 120

    Standardization of IAM Policies on Different Cloud Platforms 120

    Monitoring and Auditing Cloud Access Practices 121

    Secure Service Accounts and Workload Identities 121

    Emergency Access Planning Activities 121

    Automating IAM Compliance Checks 122

III. Challenges and Solutions in Managing Multicloud Access 122

    Identity Fragmentation and Siloed Access 123

    Inconsistent Permission Models 123

    Lack of Centralized Auditing and Compliance 125

    Managing Service Accounts and Machine Identities 125

    Compliance on Various Jurisdictions 126

IV. Emerging Trend: The Role of Zero-Trust Architecture (ZTA) in Hybrid Cloud Environments 127

    Key Strategies to Use 128

    References 129

Chapter 9: Securing Sensitive Data with IAM 131

I. Managing Access to BigQuery, Cloud Storage, and Spanner 131

    Fundamentals in IAM for BigQuery, Cloud Storage, and Spanner 131

    Handling Access to BigQuery 132

    Handling Access to Cloud Storage 134

    Handling Access to Cloud Spanner 136

    Auditing and Monitoring Access 138

II. Using Customer-Managed Encryption Keys (CMEK) for Data Security in Google Cloud 138

    Implementing CMEK Across Key GCP Services 139

    Best Practices for Managing CMEK 140

III. Fine-Grained Access Control for Sensitive Resources 141

    Best Practices for the Implementation of Fine-Grained Access 142

IV. Industry-Specific Use Case: Protecting Healthcare Records with CMEK and IAM............... 143

References 144

Chapter 10: AI-Driven Identity and Access Management 147

I. Introduction to AI in IAM 147

    AI in Cloud Security 147

    Benefits of Using AI in IAM 148

    AI-Driven IAM Tools in GCP 149

II. Anomaly Detection with AI 149

    Using Policy Troubleshooter and Cloud Audit Logs with AI 150

III. Automating Policy Management with AI 151

    AI-Based Policy Recommendations in GCP 152

    Predicting Least Privilege Access Using Historical Data 153

    Real-World Example: Automating Role Assignments in a Multi Project Organization 153

    Implementing AI-Driven Policy Management 154

IV. AI-Powered Identity Verification 155

    Leveraging AI Tools for Biometric Verification 155

    Automating Identity Verification Workflows 156

    Use Case: Implementation of AI-Enhanced Multifactor Authentication 156

V. Ethical Considerations in AI-Driven IAM 157

    Addressing Bias in AI Models for Identity Management 157

    Ensuring Transparency and Fairness in AI-Driven Access Control 157

    Aligning AI with Compliance Standards 158

VI. Emerging Trends and Future Directions 158

    Predictive IAM: Proactive Access Basing on Behavioral Patterns 158

    AI-Enhanced Zero-trust Frameworks 159

    Real-World Case Study: Using AI-Driven IAM at a Global Enterprise 161

References 161

Index 163

About the Authors

Advait Patel is a skilled senior site reliability engineer based in Chicago, with a passion for leveraging technology to drive impactful solutions. With extensive experience in cloud computing, cloud security, and cybersecurity, he currently works at Broadcom, where he plays a key role in managing, building, and securing multimillion-dollar revenue-generating products. Advait is also an advocate for professional growth and is eager to share his expertise with the next generation of tech talent through community involvement and mentorship. In his free time, he enjoys connecting with like-minded professionals and exploring innovative developments in the tech industry.

Saai Krishnan Udayakumar is a seasoned software engineer and cybersecurity practitioner based in Seattle, with more than a decade of experience building secure, scalable identity and access management systems. At a Fortune 500 enterprise, he leads critical initiatives in access governance, compliance automation, and platform security—safeguarding large-scale, high-impact infrastructure. Saai actively contributes to the broader tech community through his involvement in professional societies, technical peer reviews, and global speaking engagements. Passionate about bridging security, engineering, and operations, he enjoys mentoring emerging talent and exploring advancements in identity and cloud security.

ABOUT THE AUTHORS

Hariharan Ragothaman is a seasoned senior software engineer with more than a decade of experience architecting scalable systems across embedded platforms, cloud-native environments, and AI-powered infrastructure. At Advanced Micro Devices (AMD), he leads mission-critical initiatives in data center validation and server-side systems. Previously, he contributed to DevSecOps transformations at Athenahealth and played a key role in launching flagship consumer audio products at Bose Corporation.

He earned his MS in electrical and computer engineering at Northeastern University and his BE at Anna University. Hariharan thrives at the intersection of embedded software, distributed systems, robotics, and DevSecOps—solving high-impact technical challenges through principled system design and execution.

In parallel with his industry work, Hariharan actively contributes to academic and research communities. He has authored peer-reviewed publications in leading IEEE conferences and journals, with interests spanning edge computing, secure AI systems, and resilient infrastructure. He reviews for top-tier venues including AAAI, NeurIPS, ICLR, ICML, IJCAI, IJCNN, iLRN, SoftwareX, and the Journal of Open-Source Software. A recognized thought leader, Hariharan has delivered talks at ACM, OWASP, Conf42, SwampUP, and IEEE events. He frequently serves as a mentor and judge at global hackathons hosted by MIT, UC Berkeley, and Boston University.

His work has earned several accolades, including the Global InfoSec Award for Cybersecurity Identity Governance at RSA Conference 2025, multiple Awards of Excellence at Bose, and induction into the Athenahealth Hall of Fame.

About the Technical Reviewer

Aparna Achanta is a security architect and leader at IBM Consulting with extensive experience driving mission-critical cybersecurity initiatives, particularly in federal agencies. She has successfully implemented cybersecurity frameworks like zero trust and security by design for federal clients, strengthening the security posture and enhancing data protection and security standards across cloud applications. Her leadership has resulted in the establishment of security review boards, secure development practices, vendor evaluations, threat modeling, vulnerability management, code scanning, observability, and performance monitoring to ensure that enterprise comply with stringent federal guidelines.

Aparna specializes in securing emerging technologies in federal agencies, including low-code, no-code applications, and generative AI applications.

Aparna is a member of the Forbes Technology Council. She shares valuable insights on generative AI security challenges, contributing to industry discourse. A passionate advocate for women in tech, Aparna is a founding member and speaker at the WomenTech Network, where she inspires and empowers more than 140,000 women with talks on cybersecurity career growth. She is also an executive board member at the Austin chapter of Women in Cybersecurity (WiCyS) and an advisory board member at George Mason University's Center for Excellence in Government Cybersecurity Risk Management and Resilience.

Aparna was named to the list of 40 Under 40 in Cybersecurity by TopCyber News Magazine for her contributions to the cybersecurity community.

Acknowledgments

We would like to express our sincere gratitude to **Aditee** and **Nirmal** from the Springer Nature publication team for their invaluable support, patience, and guidance throughout the publishing process. Their attention to detail, responsiveness, and encouragement made this journey both smooth and rewarding.

A heartfelt thank-you to **Julie Ann Davis** for her unwavering support and encouragement, which played a key role in bringing this book to life.

We are especially grateful to **Aparna Achanta** for her thoughtful and constructive review of the manuscript. Her insights helped refine our ideas and elevate the clarity and impact of this book.

To our families, friends, and mentors—thank you for the love, inspiration, and belief in us throughout this journey. Your presence has made all the difference.

And to **Beacon**, for reminding us to pause, play, and be present.

Introduction

Security is no longer a background process; it's a strategic priority in a world where digital transformation is what sets businesses apart. Companies in every field are moving to the cloud to grow quickly, but this raises an important question: who should be able to see what, and how do we make sure it's safe?

Identity and Access Management (IAM) on Google Cloud Platform is where this book starts to tackle this problem. IAM is more than just a technical standard; it's the plan for building trust online. If you work in IT, security, or cloud architecture, you need to know about IAM to protect your infrastructure, make sure you're following the rules, and let innovation happen without compromising security.

We explain the basic ideas behind IAM, look at its most important parts, and talk about how it is becoming more important in a wide range of real-world fields, from healthcare to fintech. You'll learn about how IAM affects cloud strategies today and why GCP is a strong, flexible platform for putting them into action.

Let's look at the "why" behind the "how" before we get into policy settings and command-line tools. In the world of cloud security, knowing the lay of the land is just as important as getting around it.

CHAPTER 1

Introduction to Identity and Access Management on Google Cloud Platform

Overview of Identity and Access Management

Identity and Access Management (IAM) contains policies, technologies, and processes that enable organizations to decide who has the right to access resources at different points in the organization. This ensures that the resources in the organization are protected from unauthorized access. Notably, IAM plays an instrumental function in handling user identities, ensuring access controls, and complying with security regulations in contemporary organizations. IAM has key components that help in administering the organization, as summarized in Figure 1-1.

Key components of IAM include the following:

- **Access management:** IAM ensures that every user can access only the resources that they are authorized to have. The access control policies like role-based access control (RBAC) and attribute access control (ABAC) help to siphon whoever has access to whatever resource on the system, enabling an increased capacity to manage whatever needs are on the system. Additionally, mechanisms that can be used on access management include multifactor authentication (MFA), single sign-on (SSO), and session management.

CHAPTER 1 INTRODUCTION TO IDENTITY AND ACCESS MANAGEMENT ON GOOGLE CLOUD PLATFORM

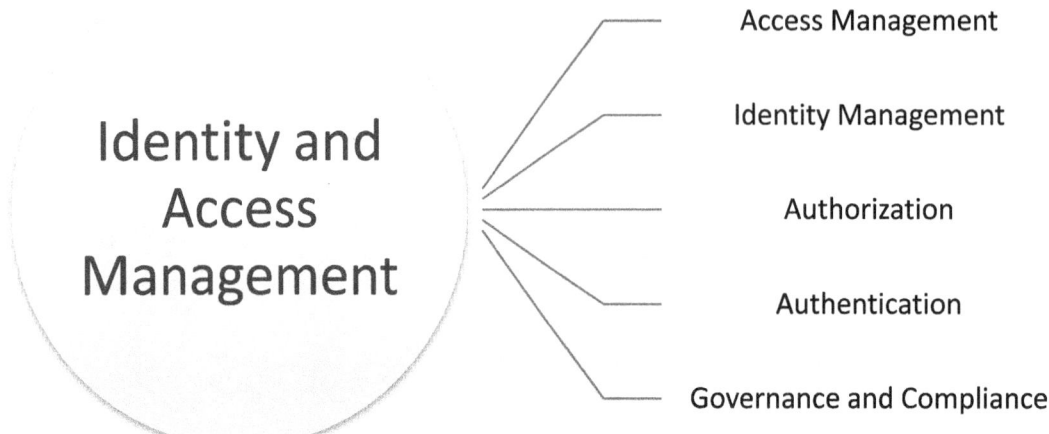

Figure 1-1. Components of IAM

- **Identity management:** This is a critical component that helps create user identities. The platform helps with profile management, user registration, and the authentication process to advance the needs of the system (Ghaffari et al., 2022). Thus, identity management ensures that every user has a unique identifier that can help in addressing whatever devices, users, and services can be handled in the organization.

- **Authorization:** This is a key component that determines the actions a user can carry out on the organizational resources. This specifies the access control list policies for the users. Therefore, users have to work based on identities that design and administer critical frameworks of desired components.

- **Authentication:** This component helps in verifying user identities and making sure that users can access a resource. Authentication can be in the form of passwords, tokens, biometrics, and certificates. These methods assist in appropriately managing the platforms (Carnley & Kettani, 2019).

- **Governance and compliance:** This component helps enforce organizational policies. The component helps to affirm access reviews and audit trails mechanisms of reporting. The framework ensures successful modeling of the IAM within the company.

These components enable IAM in any organization, ensuring the system can facilitate privacy, safety, and functionality in managing the organization. In other words, IAM helps create successful outcomes within the company when identifying and authorizing users onto the network.

Benefits of IAM

IAM has various benefits for an organization that needs to manage various users. Using IAM in an organization offers the following benefits, as summarized by Figure 1-2:

- **Operational efficiency:** IAM simplifies user onboarding. This also makes it easier to offboard users from a project or the organizational system.

- **Enhanced security:** Using IAM reduces risks due to breaches and insider threats. It keeps information and data within the company safe. Therefore, unauthorized access to sensitive resources and data within the company is well guarded (Cremonezi et al., 2020).

- **Regulatory compliance:** Using IAM makes it easier to comply with regulations such as GDPR, ISO 27001, and HIPAA. Using IAM enables the organization to audit their records more easily.

- **User experience:** IAM provides agility and access to the system without having to reduce security. This makes the user experience convenient.

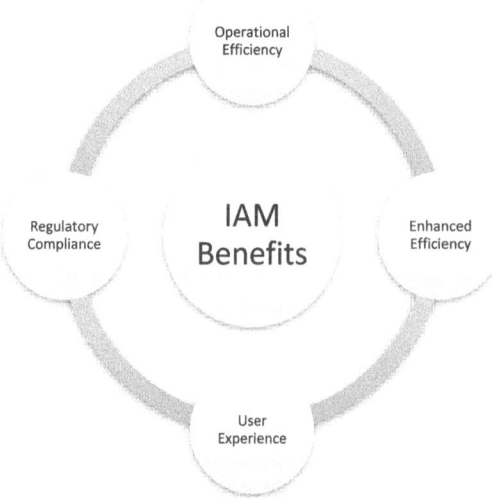

Figure 1-2. *Benefits of IAM*

IAM in Modern Organizational Environments

In modern organizations, IAM ensures that technologies and approaches can achieve sustainable results. IAM can administer relevant outcomes at all levels. The use of technologies such as zero-trust architecture ensures that IAM has the chance to create the right verification and minimal privilege that assists in handling IAM demands at all times. The modern organization can also use cloud IAM to manage identities and access in cloud-based services. IAM has also been integrated into microservices and DevOps, enabling companies to create automated workflows.

Future Directions of IAM

The development of IAM faces considerable advances when working with AI. Integrating AI with IAM ensures an identity lifecycle management, anomaly detection, and predictive access control, which enhances the safety and security standards of IAM approaches. Nonetheless, decentralized identity in IAM will help provide users with control over digital identities and make traceability of information much easier to configure and integrate within the system (Jasper et al., 2023). Adaptive access control creates the best platform for working with context-aware decisions in the system;

crafting the best scope for device, location, and user behavior; and enhancing the facilitation of identity to a great new level. Thus, IAM can be handled to achieve much better assignments in an organization and ensuring safety at all levels.

Overview of IAM on GCP

IAM on GCP consists of various identities, resources, and policies that enable the successful management of the platform. Using IAM on GCP demands a look at these elements.

Identities

There are different identities of IAM on GCP to assist in managing the organization. Some key identities in IAM on GCP includes:

- **Service Accounts:** This is an identity that handles the accounts that are used by virtual machines, applications, and other services. They are used to interact, associate, and enable authentication with other GCP resources by managing credentials in applications. In most instances, the service accounts do not have credentials that enable them to get into the platforms.

- **Groups:** Users on GCP can be organized into groups where they can have a collective permission. Google groups ensure that the administrator can provide a high level of abstraction for user permissions, handling assigning permissions, creating tasks, and even handling authentications, as detailed in Figure 1-3.

- **Users:** These are singular accounts that have been linked to Google accounts. Users will have different roles such as partners, team members, and even contractors. Each user needs access to different resources on the cloud platform (Mohammed, 2019).

CHAPTER 1 INTRODUCTION TO IDENTITY AND ACCESS MANAGEMENT ON GOOGLE CLOUD PLATFORM

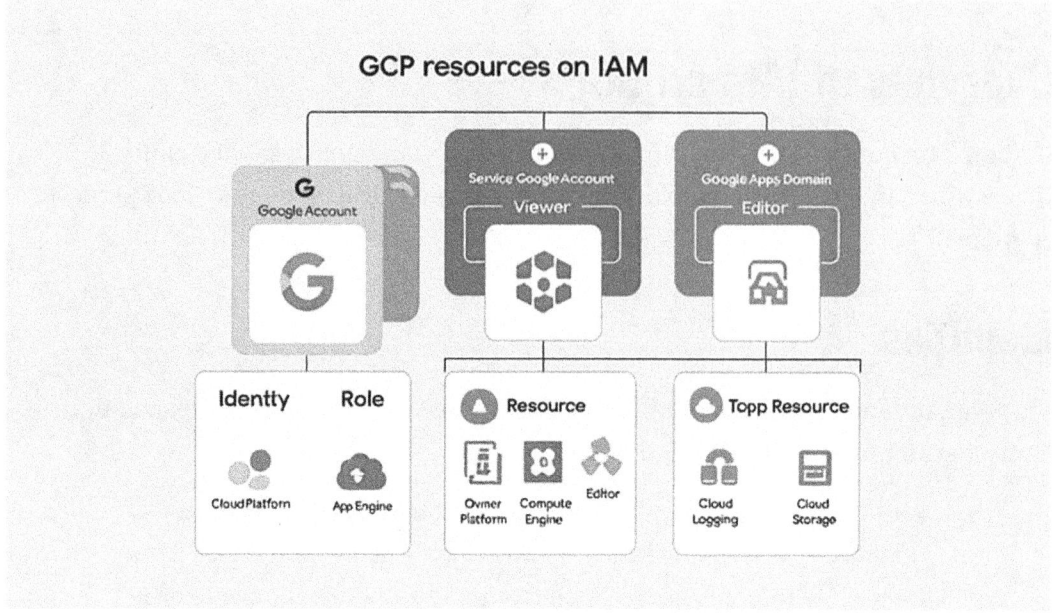

Figure 1-3. GCP resources for IAM

Resources

Resources are entities that have to be managed within the GCP. They are different components that help facilitate activities and roles on GCP, as depicted in Figure 1-3. The following are key resources within GCP:

- Kubernetes clusters
- Cloud storage buckets
- Compute engine instances
- Pub/Sub topics
- BigQuesty datasets

These resources have different levels and proportions within GCP.

Roles

Roles are assigned within the IAM system to enable a single user to carry out different activities. The roles have to be precise in detailing the right category that is needed. They have to be assigned based on the capacity and responsibilities that a user has within the system. These roles include the following capacities:

- **Custom roles:** These roles cater to users that have unique capabilities. The administrators have to create these roles. The custom roles help in fulfilling special capacities that are not already defined (Glocker et al., 2024).

- **Predefined roles:** These roles have already been created for particular GCP services based on key roles. Each of these roles has already defined their duties. An example is Storage Object Viewer, which gives read-only access to files and objects within the GCP cloud storage.

- **Basic roles:** These are roles that are provided to address basic organizational needs and activities. These roles stipulate the needed entities in the system. The roles include broad categories such as Editor, Viewer, and Owner, as indicated in Figure 1-3. However, these roles do not have any specific categorization, which helps address the principle of least privilege.

Policies

Policies are regulations and stipulated approaches that help bind identities to particular resources and roles. These policies enable the development of a hierarchy of organizational activities (Cai et al., 2019). The policies are thus key to developing and acquiring instructional approaches to meet the needs of every party, as shown in Figure 1-3. The permissions are given at the broad organizational level, flowing down to individual aspects such as projects, folders, and individual projects, making sure that they can be handled and executed within the desired timeline.

Real-World Significance of IAM

IAM is used within contemporary companies to meet organizational needs. IAM enables companies to provide authentication, access, and authorization with much ease, facilitating individual rights on the system. Various industries such as fintech, healthcare, and gaming have used IAM.

IAM Use in Fintech: Fostering Security and Compliance

The fintech industry actively uses IAM solutions to help with their user management. Because the fintech industry involves the use of highly sensitive personal and financial data, organizations need to protect data from malicious criminals. IAM helps companies with fraud prevention, regulatory compliance, and robust security advances within the industry. It helps in advancing protection to the sector through the following measures:

- **Secure transactions:** Using IAM ensures that fintech platforms can authenticate individual users and authorize their transactions. Multifactor authentication (MFA) ensures that there is a way to identify the user through passwords and tokens on devices, as shown in Figure 1-4. Additional methods such as biometrics assist in enabling user access to the platforms (Daah et al., 2024). For example, PayPal uses IAM to ensure users can authenticate their transactions and give access.

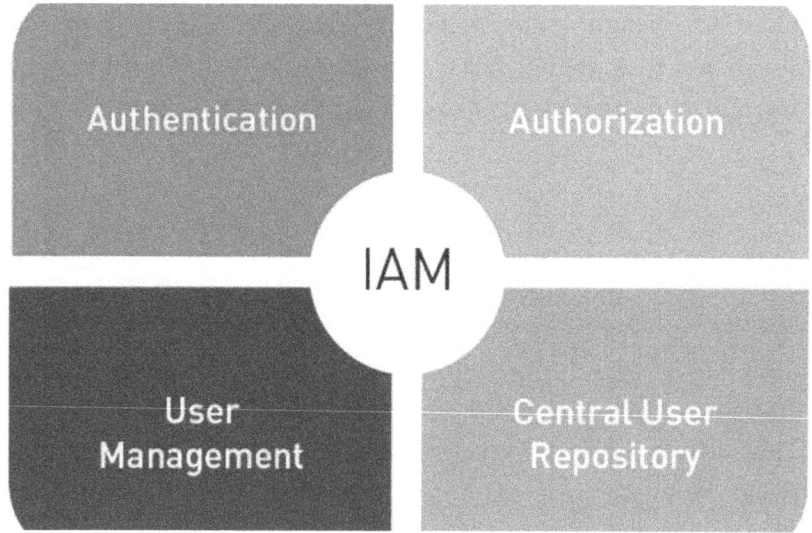

Figure 1-4. *IAM uses in the fintech industry*

- **Regulatory compliance:** Fintech companies have to comply with data security demands, so IAM helps them align with legal considerations. Regulations such as GDPR, PCI DSS, and AML have to be addressed within apps to ensure that users can authorize their data. For example, Stripe use IAM to ensure access to customer data, aligning with global requirements and regulations on the management of customer information.

- **Fraud prevention:** IAM employs artificial intelligence (AI) and machine learning (ML) to help manage transactions and prevent fraudulent activities. AI-powered IAM activities can detect anomalies, making it much simpler to access protection for the customer accounts, as detailed in Figure 1-4. This helps banks address customer safety, detect anomalies, and prevent fraud. For instance, mobile banking applications block the transfer of huge amounts of money on new devices, when there is no prior authentication for the transfers.

Use of IAM in Healthcare to Protect Patient Data and Enable Continued Access

Advancements in the healthcare sector have called for the use of technology-based approaches in managing its privacy and access needs. Legal requirements to protect patient data demand the use of IAM and related features to advance solutions to the community at all levels. IAM can handle usability, compliance, and security at all levels. IAM works with the healthcare sector in the following ways:

- **Streamlining access for providers:** Healthcare providers work with IAM to ensure that they can get immediate access to patient data in cases of emergencies. The use of single sign-on ensures that administrators and practitioners have access to several systems with a single set of credentials.

- **Patient privacy:** Electronic health records (EHRs) store patient data, calling for more protection mechanisms to help address the increased amount of data stored on the platforms (Lawson, 2020). IAM enables only authorized parties to access these records, maintaining privacy, as shown in Figure 1-5. Companies like Epic Systems, for example, ensure that its EHR systems uses IAM to restrict access based on individual attributes such as role and location.

Figure 1-5. IAM use in healthcare industry

- **Regulatory compliance:** Organizations have to manage data in accordance with regulations such as HIPAA and GDPR. The use of IAM ensures that there are access controls, encryption, and audit trails when handling information, as depicted in Figure 1-5. These advances craft the best scope and chance of channeling value to the healthcare institutions while maintaining the best level of safeguarding information.

IAM in Gaming to Enhance User Experience and Address Cheating

The gaming industry uses seamless and intriguing user experiences to engage users. IAM offers the capacity to handle secure access, fair play, and fraud. Online gaming uses IAM in the following ways:

- **Seamless user experience:** IAM enables the user of SSO and identity management in federated systems so players can have the same credentials across games. For example, Xbox uses IAM so users can enjoy gaming on several consoles, mobile devices, and PCs, without having to create multiple accounts.

- **Fraud prevention:** IAM makes the use of advanced analytics much easier. It integrates analytical data to help highlight fraudulent behavior. The use of IAM makes it much simpler for gaming companies to flag suspicious activities and send alerts to the right party.

- **Account security:** Gaming companies use models such as biometric authentication and MFA that assure users' safety (Truong, 2021).

IAM is used in the real world for everything from access control and identity management. For example, healthcare uses IAM to handle patient privacy and ensure access to providers even in emergency situations. Fintech uses IAM for preventing fraud and handling regulatory compliance. The gaming industry uses IAM to make it easier to handle cheating and provide a seamless user experience through single sign-on. These techniques enable growth in organizations and help manage safety in the cyber world.

Tools Required

Configuring IAM on GCP requires various tools:

- **Gcloud CLI:** Gcloud CLI assists with command-line functionalities. GCP uses Gcloud CLI to help manage cloud resources. The approach enables developers and administrators to work with GCP services to ensure automation, integration, and scripting of IAM tasks within the CI/CD pipelines. Gcloud CLI assists in handling policies, service accounts, and roles for several projects and resources. Also, Gcloud CLI can assist in the process of managing permissions by granting or revoking them when needed.

- **Google Cloud Console:** The Google Cloud Console offers a user-friendly platform for handling IAM roles, permissions, and policies. It helps in managing the audit process for logs, looking into every IAM actions, and ensuring that policy changes and role assignments can be conducted. Nonetheless, the platform offers visual insight into whatever actions and elements can be conducted within the software management service (Roy et al., 2021).

- **APIs:** GCP works with several APIs to help with the management of IAM. Key APIs are Resource Manager API, Service Account Credentials API, and IAM API (Kingsley, 2023). These APIs can help developers build custom tools to grant permissions, create automated workflows, and assign roles.

- **Terraform:** This is an optional tool that can be used in the GCP platform to ensure that there can be an infrastructure as code (IaC) to help manage roles and policies on GCP. Terraform can define IAM resources within configuration files and make the application consistent within different environments. Terraform can be used in versioning, ensuring that the configurations are stored and are reproducible in the course of managing IAM.

Leveraging the use of these tools is key to advancing the functionality of IAM on GCP. Table 1-1 summarizes these tools and their function.

Table 1-1. Tools for IAM on GCP

Tool	Purpose	Use Case
Gcloud CLI	Command-line platform key in managing service accounts, roles and IAM policies	CI/CD integration Scripting Automation
Google Console	IAM management using web interface	Visualization Manual management
APIs	Accessing further IAM functionalities	Integration of applications Customization
Terraform	IAM management	Infrastructure deployment

Emerging Trends

The evolution of technology has led to the creation of various platforms seeking to converge with IAM to enhance security. Technological advancements in areas such as zero-trust principles and access control with AI have created changes for IAM. This collaboration can result in greater safety and security. Key emerging trends include the following:

- **AI-driven access control:** The development of AI and ML has enhanced IAM by bringing in intelligence and proactive and adaptive access control approaches. The IAM models ensure that there are key advances in modeling the policies and rules, regulating advancement, and crafting the best capacity to achieve stable advancement in all levels. Therefore, AI-driven access control works with key advances such as behavioral analytics, which helps to flag suspicious activities and call for more authentication models. Nonetheless, risk-based authentication enabled by AI ensures that there are different levels of authentication based on the risk of a single request. AI also enables predictive threat detection, achieving risk mitigation and prevention of breaches (Sarker et al., 2021). AI also automates policy enforcement. Thus, AI is a chance to get more functionality with access control.

- **Zero-trust principles in IAM:** Zero-trust security means there is a constant demand to verify one's identity on the network. This indicates no trust for any device or user within the platform. Zero trust works based on continuous verification and least privilege access, which ensures that there are micro segments for users, enabling them to consistently address their needs without having to gain information that they do not require (Garcia-Teodoro et al., 2022). These principles are the best way to handle IAM through categorizing provisional advancement of information to users on the networks. In IAM, it helps by ensuring real-time monitoring for the logins and user activities on the platform. Identity-centric security models handle the security of the community. Using context-aware policies also ensures that the user is within the right location before giving out information.

In summary, AI and zero trust work with IAM to help advance security and identity. The use of AI ensures zero trust can use real-time and context-aware approaches to handle access. Thus, these emerging technologies improve IAM on GCP.

References

Cai, F., Zhu, N., He, J., Mu, P., Li, W., & Yu, Y. (2019). Survey of access control models and technologies for cloud computing. *Cluster Computing, 22*, 6111-6122.

Carnley, P. R., & Kettani, H. (2019). Identity and access management for the internet of things. *International Journal of Future Computer and Communication, 8*(4), 129-133.

Cremonezi, B., Vieira, A., Nacif, J. A., & Nogueira, M. (2020). Survey on identity and access management for internet of things.

Daah, C., Qureshi, A., Awan, I., Adalat, O., & Konur, S. (2024, August). Advancing IAM in the Finance Sector by Integrating Zero Trust and Blockchain Technology. In *International Conference on Mobile Web and Intelligent Information Systems* (pp. 83-99). Cham: Springer Nature Switzerland.

García-Teodoro, P., Camacho, J., Maciá-Fernández, G., Gómez-Hernández, J. A., & López-Marín, V. J. (2022). A novel zero-trust network access control scheme based on the security profile of devices and users. *Computer Networks, 212*, 109068.

Ghaffari, F., Gilani, K., Bertin, E., & Crespi, N. (2022). Identity and access management using distributed ledger technology: A survey. *International Journal of Network Management, 32*(2), e2180.

Glöckler, J., Sedlmeir, J., Frank, M., & Fridgen, G. (2024). A systematic review of identity and access management requirements in enterprises and potential contributions of self-sovereign identity. *Business & Information Systems Engineering, 66*(4), 421-440.

Jasper, K. D., Raja, A. V., Neha, R., Rajest, S. S., Regin, R., & Senapati, B. (2023). Secure Identity: A Comprehensive Approach to Identity and Access Management. *FMDB Transactions on Sustainable Computing Systems, 1*(4), 171-189.

Kingsley, M. S. (2023). Google Cloud Platform (GCP) Lab. In *Cloud Technologies and Services: Theoretical Concepts and Practical Applications* (pp. 325-378). Cham: Springer International Publishing.

Lawson, H. (2020). AI-Driven Multi-Factor Authentication: Enhancing IAM Security in Healthcare Systems.

Mohammed, I. A. (2019). Cloud identity and access management–a model proposal. *International Journal of Innovations in Engineering Research and Technology, 6*(10), 1-8.

Roy, A., Banerjee, A., & Bhardwaj, N. (2021). A Study on Google Cloud Platform (GCP) and Its Security. *Machine Learning Techniques and Analytics for Cloud Security*, 313-338.

Sarker, I. H., Furhad, M. H., & Nowrozy, R. (2021). Ai-driven cybersecurity: an overview, security intelligence modeling and research directions. *SN Computer Science*, *2*(3), 173.

Truong, D. (2021). *Evaluating Cloud-Based Gaming Solutions* (Doctoral dissertation).

CHAPTER 2

Managing Roles, Policies, and Permissions

Types of Roles

Google Cloud Platform (GCP) allows for the use of various types of roles within its Identity and Access Management (IAM) platform. IAM encourages role-based access control so organizations can secure their databases with scalable approaches that assist in targeting system roles and data protection. Key roles include predefined roles, basic roles, and custom roles. Understanding each of these roles is key to addressing IAM within GCP and creating roles in every system.

Predefined Roles

Predefined roles in GCP come with a list of specific services that a user can have granular access to. The GCP-provided roles help users manage workflows and job functions. With the provisioning of these roles, each user can be assigned a role and automatically do particular tasks because they have the right permissions within the system (Qiu et al., 2020). Predefined roles have various features enhancing their safety and security mechanisms. Key features of predefined roles include the following:

- **Granularity:** Predefined roles have the capacity to break down permissions into actions. Keywords used for the roles include admin, read, delete and write, as indicated in Figure 2-1. This approach reduces the risk of providing more permissions than needed and enables a user to have only the permissions that closely relate to their work (Ravidas et al., 2019).

CHAPTER 2 MANAGING ROLES, POLICIES, AND PERMISSIONS

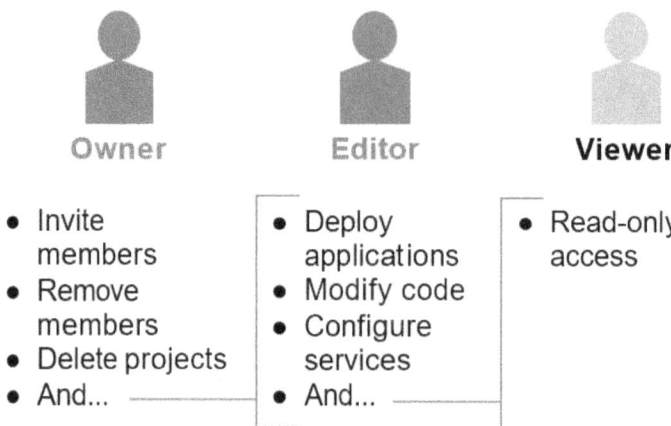

Figure 2-1. Predefined roles

- **Service-specific permissions:** Predefined roles ensure that the permissions provided to users closely relate to the specific services that they need. Services such as Compute Engine and Cloud Storage are tied to key roles, ensuring that the user does not get permission to other tasks that might affect key functionalities within the system, as detailed in Figure 2-1.

- **Updates by Google:** Google ensures that the predefined roles can be updated to enable broad management of the system. Updates in terms of security improvements, features, and categories ensure that roles can accommodate advances and that the organization can evolve alongside changing and advancing technological features.

Predefined roles can be used in specific instances, ensuring that there is a critical understanding of the core needs of the system. The roles can be handled by working on role segregation, where the delegation of duties and management activities helps to avoid unsolicited access. This ensures management is separated from network needs, facilitating a flawless handling of the administrative activities. Predefined roles also work alongside the least privilege principle, enabling a user to have access to the right category of tools that will help them achieve their goal on the system without anything extra.

Custom Roles

Custom roles have the advantage of flexibility and uniqueness from one organization to the next. They ensure that organizations can combine different permissions into a single role, helping them manage an organization's complex workflow. These custom roles assist in achieving valuable results for the system's advancement needs. Regardless of the industry, organizations have security demands and can use custom roles to comply with regulations. Custom roles have the following features:

- **Static and predictable:** Custom roles have a static capacity, where the organization sets them once. The administrator provides an outline of the custom activities to be conducted.

- **Service specific:** Custom roles can have permissions that relate to a single service or that handle several services to achieve the same functionality. The roles can be modified to cater to as many services as possible (Indu et al., 2019).

- **User-defined permissions:** GCP provides a list of permissions that administrators can use for specific identities and duties on the platform, increasing the scope and capacity to address the different duties of users.

Custom roles can be used in different instances. Organizations can create roles that combine several services, ensuring the use of multiple services through a single permission. The organization can provision an increased scope of administering duties to a defined level. Also, custom roles can eliminate unnecessary permissions on the system, ensuring that redundancy is removed, by having a single role that encompasses multiple functions. Additionally, duties can be segregated, which is a beneficial way to achieve sustainable results when dealing with roles and duty execution in the system.

To use custom roles, there are different practices that an organization needs to include in their management approach. First, roles can be created from predefined roles. Having a reference point in the predefined roles ensures that essential permissions and duties within the system are covered and can assist in targeting increased service development. Second, organizations have to constantly look into permissions to ensure that changing needs within an organization have the right scope and level. This approach will enable roles to be evaluated, ensuring that they do not overlap or have a negative result.

CHAPTER 2　MANAGING ROLES, POLICIES, AND PERMISSIONS

Basic Roles

Basic roles are broad and can be referred to as *legacy roles*. They are coarse grained, ensuring that they can cover a wide enough area for individual users to handle their duties as much as possible (Alsirhani et al., 2022). The broad nature of these roles ensures that they can be used in cases where an organization has few employees. The key features of these roles include the following aspects:

- **Broad permissions:** These roles can be categorized into three distinct types: Owner, Viewer, and Editor. The categorization enables a strict understanding and management of the role handling within the system, as indicated in Figure 2-2.

Figure 2-2. Broad permissions on IAM

- **Global application:** The roles are defined to handle project-level access, ensuring they advance suitable results whenever desired by the organization.

Basic roles can be applied in different instances, ensuring a user can access information and data on different projects on a temporary basis. These roles also have a quick onboarding approach, so they are good to use in testing phases to quickly manage basic data and execute roles.

Despite the ease of use provided by the basic roles, they have various limitations that prevent them from being used in full-scale activities. The over-permissioning aspect of the roles affects the privacy of sensitive data. Basic roles also fail to comply with regulatory requirements for handling information, further leading to a need to identify use cases and potential negative effects when using basic roles.

Best Practices in Role Management in GCP IAM

To address role management in GCP, different approaches can be used. The key steps in role management include the following aspects:

- **Monitor and audit roles regularly:** Tools such as IAM Policy Analyzer help with the review of access points for every user. This creates the chance to manage roles and advance critical safety approaches.

- **Principle of least privilege:** Having minimum permissions is key to ensuring that users do not violate their access to the information and that they have the right data needed to complete their tasks.

- **Use folders and projects for isolation:** Roles and folders have to be applied within any project to ensure security boundaries are well administered and that users are able to achieve the desired results (Ghadge, 2024).

- **Leverage predefined roles:** Managing roles has to begin with predefined roles, as they offer the right category of security and safety for a system.

IAM Policies and Inheritance in the GCP Resource Hierarchy

GCP uses IAM to ensure that administrators can specify each user that has access to each resource on the platform. The use of the IAM policies ensures that there are specific permissions for groups and users for the resources that they have to use. Policies are instrumental in addressing the GCP resource hierarchy, advancing inheritance to assist in streamlining, handling access management, and working within organizational environments to achieve designed outcomes (Kern et al., 2023). Thus, using IAM policies is key to ensuring that permission structures are scalable, secure, and efficient.

CHAPTER 2   MANAGING ROLES, POLICIES, AND PERMISSIONS

IAM Policies in GCP

IAM policies are roles and bindings that define the roles and whatever they have to execute within the system. Policies specify who can be granted permission within the system, such as a user, group, or service account. Policies define the roles, such as the identity and permissions, that one has within the system. They also define the scope of what users can carry out on the system.

IAM policies have different features for identifying and managing individual responsibilities. The IAM policies have features like the following:

- **Roles:** These are permissions categorized together within the system. They can be basic, custom, or predefined, enabling a user to carry out particular tasks in the system (Talluri, 2023).

- **Principle:** This is an identity provided to policies to specify critical engagement dynamics.

- **Conditions:** These are rules that enable policies to offer permissions in the specific order that is outlined. They ensure that there is a way to handle access and location, helping to enforce security in the framework.

Using policies ensures an increased capacity to address and manage critical components of the system.

GCP Resource Hierarchy

GCP has an organized approach to managing resources. The hierarchical structure consists of four primary levels that each define, address, and mitigate the key features of the system. These levels are as follows:

- **Organization:** This is the root, encompassing the entire company. It ensures that the company domain is well addressed and represents the organization in full, as shown in Figure 2-3.

- **Folders:** These are subdivisions that are presented within the organization. The folders can work within each other and help to understand the entire scope of activities (Roy et al., 2021).

CHAPTER 2   MANAGING ROLES, POLICIES, AND PERMISSIONS

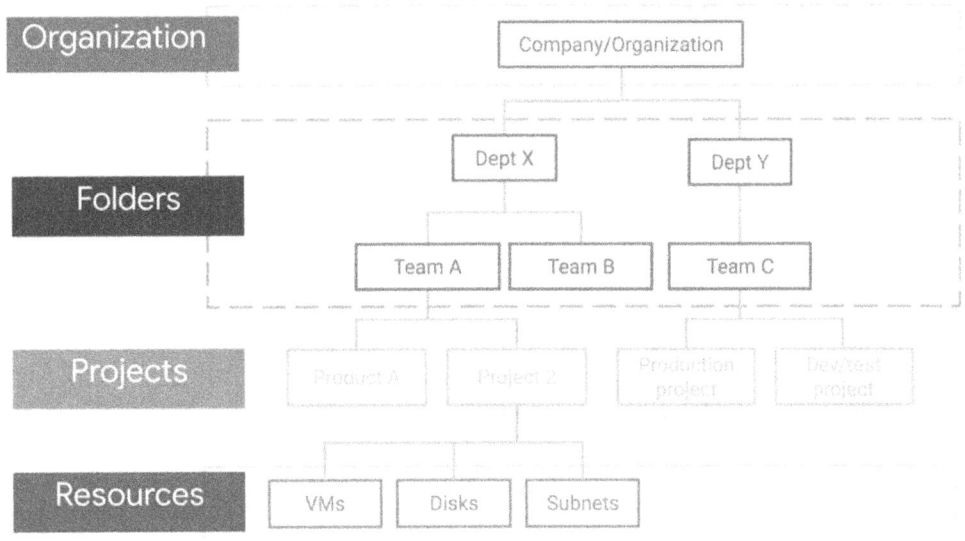

Figure 2-3. GCP resource hierarchy

- **Projects:** These are considered resources, such as Cloud Storage buckets, Compute Engine instances, and BigQuery buckets. Each project identifies a GCP resource that should be handled to achieve the desired value, as indicated in Figure 2-3.

- **Resources:** These are cloud services and objects that work in a project.

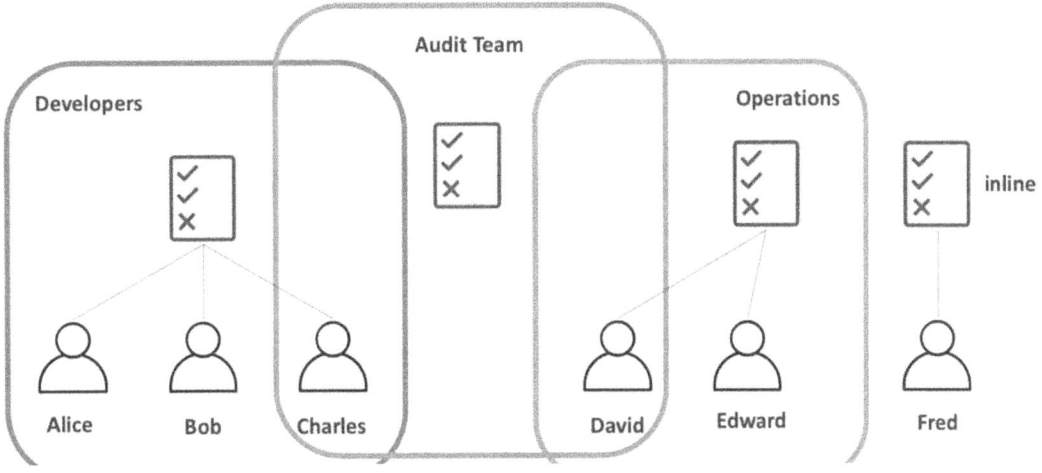

Figure 2-4. Policy inheritance

This hierarchy ensures a logical framework for handling access and modeling IAM policies. The specific levels of IAM policies ensures that an individual can inherit from the parent and meet their goals.

IAM policy inheritance

GCP IAM policies have to follow a hierarchical inheritance approach. These policies exist within the hierarchy, as they are set within the higher level and apply to any resources that come from the parent hierarchy. Therefore, inheritance can address individual needs based on the parent modeling and classification activities. Inheritance works in different ways, through different approaches such as these:

- Any policy given to the organization level applies to projects, folders, and projects that are within the organization.
- Policies in the folder level will relate to all projects and resources that exist within said folder.
- Policies in the project level work within every resource in the project, addressing every needed element from that point onward.
- Policies in the resource level impact only the resource, ensuring that there are different ways to cater to the required installations for the policies (Khambam & Kaluvakuri, 2023).

The hierarchy ensures every provision works to achieve every demand. For instance, when a user has the Viewer role at the organizational level, they have the right chance to look at every resource within the organization. A different user with a Viewer role in the storage.object Viewer capacity will be limited to looking at objects in said bucket, ensuring they can conduct their activities only within the mentioned category. The hierarchy is therefore key to defining the access that every individual has in the system.

Policy Evaluation in the Resource Hierarchy

GCP has the capacity to evaluate the actions of any user. For instance, when a user attempts to carry out an action, GCP has to evaluate permissions tied to the user. GCP starts with the resource, looking at all the policies that relate to the specific resource. If there is no policy to give access, GCP moves to the project level (Cabianca, 2024). At this level, if no permissions are found, the process continues to the folder and organization levels. Access is granted to a user if there is a policy within the hierarchy that gives them the right permission. In cases where no permission is granted within the hierarchy, GCP will deny the permission to the user.

Overriding IAM Policies

Policy management is mainly conducted using inheritance, which checks the policies and whatever capacity they can grant these policies. However, in certain instances, policies can be overridden by administrators. GCP enables administrators to override these policies by defining the policies within lower levels of the hierarchy, thus granting the same polices. Within the GCP platform, there is no provision to *deny* any regulations. This indicates that permissions have to be provided in specific levels, ensuring that the inherited permissions relate to assigned roles. Therefore, the policy management of GCP works by ensuring that roles can only be assigned and not denied within the hierarchy, an approach that helps to manage the entire system of permissions within the hierarchy.

Best Practices for Inheritance and IAM Policies

There are several best practices for detailing the right categories of policies for users at different levels. IAM policies have to be categorized in the right manner, and the following are practices to help with IAM policies and inheritance:

- **Start at the organizational level:** Having broad policies in the organizational level will ensure that there is consistent access to the resources. This helps to ensure that users can handle duties within the determined sections easily.

- **Use folders to conduct logical grouping:** This will ensure that policies within departments and teams will be categorized together and makes it easy to manage these policies.

- **Limit over-permissioning:** Administrators should have precise permissions, ensuring that there are broad permissions for the different users on the system.

- **Regular audits:** Permissions have to be regularly audited. Use the IAM Policy Analyzer to review inherited permissions (Gouglidis et al., 2023).

- **Conditional policies:** Having conditional policies within IAM will ensure additional security and safety measures. This is a better way to cater for the management of safety and provisioned mentions when addressing individual roles.

- **Document overrides**: Having critical documentation of overrides makes it simpler to avoid confusion and maintains a great audit record. This will help with policy management.

Managing policies within the hierarchy means using GCP strategically to assist in maintaining critical control over the data.

Real-World Use Case: Managing Permissions in a Multiproject Organization

Within the modern organizational environment, multiple projects can be run concurrently. GCP assists with this, ensuring their completion and meeting the demands of the projects. Nonetheless, project management depends on permissions and system settings that allow for operational efficiency, compliance, and security. Organizational regulations have to be followed, ensuring that each project is completed appropriately. Companies use permissions to ensure the correct roles are granted to the users. One such company is Spotify, which uses GCP to manage their IAM policies and create a multiproject architecture.

Spotify Challenge

Spotify is global, with their engineering teams working on several projects simultaneously. The company deals with recommender engineers, music streaming, and data analytics. Teams work with each of these functions so that customers have a working solution at all times. Spotify uses GCP to ensure the teams have the right permissions. It also uses IAM modeling to meet their needs within the least privilege principle (Sukhdeve D & Sukhdeve S., 2023). Hence, working within the company's provisions demands inclusion of GCP and IAM to ensure they can facilitate constant updates and round-the-clock integration of their demands for the customers to have stable configurations.

Solution

Spotify's demand for multiproject engagement ensured that they worked with GCP to handle the resource hierarchy and IAM policies. This helps Spotify create the structured access required for the different teams that are collaborating across the globe. The following are the key steps that ensure Spotify is getting seamless functionality and collaboration:

- **Organization-level policies:** Spotify provides organization-wide roles in central activities such as security and compliance needs. There are key users who are assigned to review and monitor the structure of resources within every provided project. Therefore, using these organization-level policies ensures that the users can have access to policies and use them as needed.

- **Folder-level segmentation:** Spotify works with projects that demand various addresses. The grouping of the projects into folders ensures that Spotify could have machine learning models to help in a single folder and recommender engines in another. IAM policies help route users to individual folders pertaining to their projects. And they help with capturing and advancing valuable information when improving their solutions (El Founti Khsim, 2024).

- **Custom roles:** Spotify works with custom roles to help manage the workflow. This engagement enables the company to ensure that specific management is customized to help workflow management within the structure of the organization. Using custom roles ensures that they can use administrative functions as they continue monitoring and handling the needed aspects of the organization.

- **Conditional policies:** Conditional policies assist in managing access-based factors. These aspects include temporary permission grants to the incident response teams. This approach helps teams handle the policies and specific project categories.

Spotify's Outcome

The company's investment in GCP and IAM enhanced their access management on multiple projects. Using the GCP resource hierarchy in multiple projects minimized risks as they increased collaboration between the teams. It also helped them comply with regulatory standards like GDPR and helped with permission management across diverse projects. Hence, using IAM, these advances helped improve access management in Spotify.

Real-World Insights: Lessons Learned from IAM Misconfigurations in Large Enterprises

Using IAM configurations is common in protecting against vulnerabilities within cloud environments. IAM can help when tackling data breaches, unauthorized access, and compliance violations. It ensures that companies have the best options for handling their diverse teams and sprawling infrastructures. The following real-world instances indicate the lessons that contemporary organizations can learn from using IAM:

- **Capital One Data Breach (2019):** The company's data breach in 2019 led to several instances of misconfiguration, where the broad permissions by the company using IAM roles led to access to sensitive data within their network segments, as indicated in Figure 2-5. This example illustrates how the use of broad roles can result in great risk for the companies and can harm data management (Khan et al., 2022). Therefore, predefined roles are key in ensuring permissions help with specific use cases and can influence every approach when crafting valuable security needs of the company.

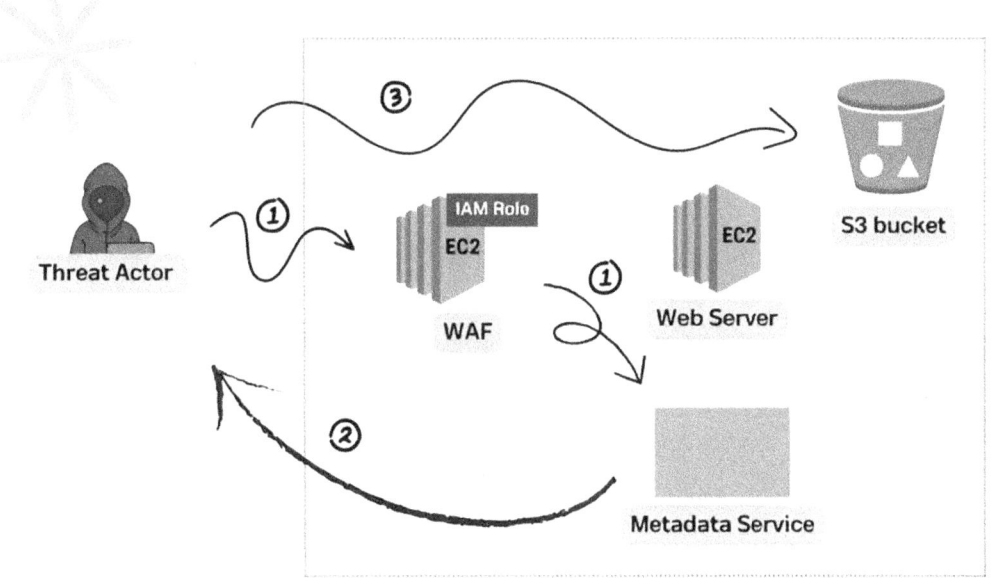

Figure 2-5. Capital One data breach

- **Uber's GitHub IAM scandal (2016):** Uber's breach occurred through access to GitHub repository that contained the company's credentials. The use of these credentials offered the attackers access to the company's sensitive resources (da Silva, 2020). The incident illustrated the importance of IAM policy audits, which will ensure that there should be a categorical instruction to look into the policies and management of every instruction. Nonetheless, the mitigation effort helped to locate the right steps in managing credentials and securing the credential handling approaches. These advances illustrated the critical nature of handling IAM credentials for any organization.

- **Snapchat's IAM policies:** Snapchat developers faced a great risk in exposing the company's sensitive data when IAM policies offered excessive permissions to the lower projects. The approach helped developers understand hierarchical inheritance, which ensures that there should be policies to restrict the access of higher hierarchies to the lower projects. This in turn enables the privacy of projects and complete security of the resources. Resource-level policies can offer the chance to accurately manage the organizational needs.

Table 2-1 summarizes these real-life lessons.

Table 2-1. Key Lessons from Real-Life Attacks

Challenge	Impact	Solution
Faulty policy inheritance regulations	Unauthorized access to sensitive information	Review inherited permissions Override the policies
Broad permissions	Unauthorized access to sensitive and unintended data	Audit permissions regularly Use the policy of least privilege
Lack of monitoring	Late detection of breaches	Set alerting system for intensive IAM activities

In conclusion, these real-world examples indicate approaches that should be used to help in enforcing IAM policies. Structuring the IAM policies to enable permission granting at specific levels, ensuring conditional policies, and working within segmented selections marks the capacity to achieve sustainable results in advancing safety within organizations. These case studies highlight the critical needs in handling GCP's IAM policies to improve security and data handling in organizations.

References

Alsirhani, A., Ezz, M., & Mostafa, A. M. (2022). Advanced Authentication Mechanisms for Identity and Access Management in Cloud Computing. *Computer Systems Science & Engineering, 43*(3).

Cabianca, D. (2024). Configuring Access. In *Google Cloud Platform (GCP) Professional Cloud Security Engineer Certification Companion: Learn and Apply Security Design Concepts to Ace the Exam* (pp. 15-175). Berkeley, CA: Apress.

da Silva, P. M. G. (2020). *Contributions to Personal Data Protection and Privacy Preservation in Cloud Environments* (Doctoral dissertation, Universidade de Coimbra (Portugal)).

El Founti Khsim, O. (2024). Empowering Start-up success: Implementing Google Cloud solutions for business growth.

Ghadge, N. (2024). Enhancing Identity Management: Best Practices for Governance and Administration. *Computer Science & Information Technology (CS & IT)*, 219-228.

Gouglidis, A., Kagia, A., & Hu, V. C. (2023). Model checking access control policies: A case study using google cloud iam. *arXiv preprint arXiv:2303.16688*.

Indu, I., Anand, P. R., & Bhaskar, V. (2018). Identity and access management in cloud environment: Mechanisms and challenges. *Engineering science and technology, an international journal*, 21(4), 574-588.

Kern, S., Baumer, T., Fuchs, L., & Pernul, G. (2023, July). Maintain High-Quality Access Control Policies: An Academic and Practice-Driven Approach. In *IFIP Annual Conference on Data and Applications Security and Privacy* (pp. 223-242). Cham: Springer Nature Switzerland.

Khambam, S. K. R., & Kaluvakuri, V. P. K. (2023). Multi-Cloud IAM Strategies For Fleet Management: Ensuring Data Security Across Platforms.

Khan, S., Kabanov, I., Hua, Y., & Madnick, S. (2022). A systematic analysis of the capital one data breach: Critical lessons learned. *ACM Transactions on Privacy and Security*, 26(1), 1-29.

Qiu, J., Tian, Z., Du, C., Zuo, Q., Su, S., & Fang, B. (2020). A survey on access control in the age of internet of things. *IEEE Internet of Things Journal*, 7(6), 4682-4696.

Ravidas, S., Lekidis, A., Paci, F., & Zannone, N. (2019). Access control in Internet-of-Things: A survey. *Journal of Network and Computer Applications*, 144, 79-101.

Roy, A., Banerjee, A., & Bhardwaj, N. (2021). A Study on Google Cloud Platform (GCP) and Its Security. *Machine Learning Techniques and Analytics for Cloud Security*, 313-338.

Sukhdeve, D. S. R., & Sukhdeve, S. S. (2023). Introduction to GCP. In *Google Cloud Platform for Data Science: A Crash Course on Big Data, Machine Learning, and Data Analytics Services* (pp. 1-9). Berkeley, CA: Apress.

Talluri, S. (2023). Saviynt Meets GCP: A Deep Dive into Integrated IAM for Modern Cloud Security. *Journal of Information Security*, 15(1), 1-14.

CHAPTER 3

Advanced IAM Features

Using IAM Conditions for Context-Aware Access

Identity and Access Management (IAM) conditions offer context-aware access control with granulated modeling. IAM conditions ensure that organizations have the capacity to identify access policies that can adapt to any contexts in which requests are made. The entire process works while ensuring that the platform is secure and flexible.

IAM Conditions

Conditions ensure administrators can provide a system of access control policies based on attributes. These fine-grained adaptations ensure there are request context, resource attributes, and identity properties. Each condition ensures that any request is established in terms of its contextual parameters; the condition considers the request's capacity and either grants access or refuses access. Some of the key attributes that have to be checked are time of day, geographic location, IP address, and device security posture (Arfoui et al., 2019). Using conditions allows developers to assign roles based on predefined capacities within the system. Additionally, IAM conditions can adapt to every request based on their specific needs. Therefore, users and services can perform certain actions only when they meet the predetermined conditions.

IAM conditions have several features, such as the following:

- **Bindings:** These serve as a critical link between the IAM roles of users and the service accounts when conditions are met.
- **Condition expressions:** These are logical statements that define how attributes in the system can be evaluated, as shown in Figure 3-1. They work to ensure that requests have been evaluated. Thus, conditional expressions assist in reviewing requests.

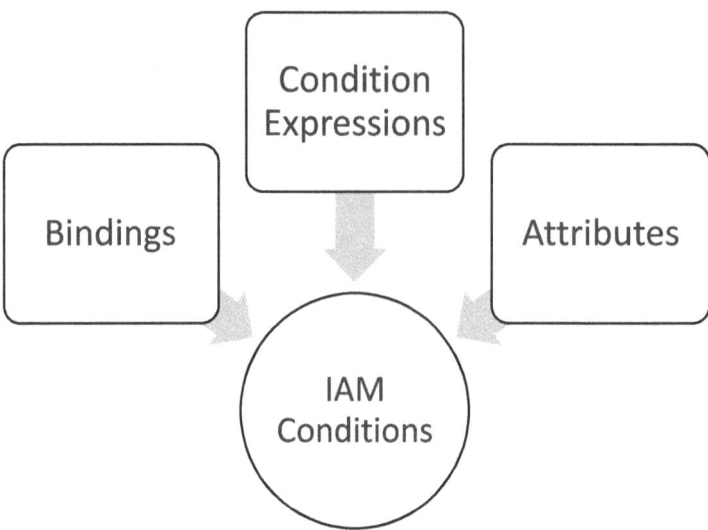

Figure 3-1. *IAM conditions*

- **Attributes:** These are contextual factors that have to be investigated for any request within the system. The attributes include identity attributes, which are the details of the service account, user, and group that are making the specific request, as shown in Figure 3-1. Request attributes include information such as the requested action, resource type, and IP address from which it originates (Qiu et al., 2020). Resource attributes consider the metadata that relates to the resource, making it simpler to learn about the resource and required advances in dealing with its requests.

IAM Conditions and Context-Aware Access

IAM conditions can refine context-aware aspects and dynamically evaluate the capacity of each access request. The approach works by ensuring policies are managed accordingly, checking every condition and making sure the requests are evaluated. IAM conditions work follow the least privilege principle, leading to using only the resources that are needed within specific circumstances. IAM conditions can considerably aid the process of context-aware access in different areas. Facilitating the context aware model includes aspects such as the following:

- **Location-based access control:** Organizational systems can be attuned to look into the locations of the requests. The companies can work by dynamically evaluating the request and ensuring that they only apply to the specific countries which they accept. The use of location-based access control makes the chance to ensure that geographic zoning can be conducted to allow handling of every scope to address and model the creation of an appropriate way to craft successful outcomes in all conditions as shown in Figure 3-2. Therefore, the use of location based attributed will help in ensuring that no request from a region outside the approved list gets to access information on the systems.

Figure 3-2. *Context-aware access*

- **Time-based access policies:** Organization administrators can define policies that include timelines for accessing data in their systems. These timelines ensure that requests will be evaluated basing on the time of day (Zheng et al., 2019). Using the approach confines employees to accessing information only during their working hours, which adds a layer of security.

- **Device security posture:** This is an attribute that considers the device that is sending a request to the network. IAM conditions ensure that compliant and secure devices have the highest chance of getting into the platforms, as indicated in Figure 3-2. The attributes from these devices have to be up-to-date, safe through encryption, and work within the organizational standards.

- **IP address restrictions:** There are specific ranges of IP addresses that can work when sending requests. Some IP addresses cannot be used to get into the institution, as the range is beyond the organization's set capacities. Essentially, this ensures that the corporate network IPs have a high priority of getting requests approved compared to external devices that will be denied access.

- **Service account and resource context:** IAM conditions offer access based on resources and services that they demand to use (Xiao et al., 2022), as shown in Figure 3-2. Therefore, using a context enables access in a single instance, which works with the user's needs.

In short, conditions secure and approve requests. The conditions, which can be used collaboratively or independently, ensure the security of requests, enabling fine and well-granulated access control on systems. The approach is key to understanding how to facilitate better safety in access control.

Benefits of Using IAM Conditions

Using IAM conditions offers several benefits for access control in an organization. Working with conditions enhances the identification and modeling of critical policies that create a safe environment. The benefits include the following:

- **Granular control:** IAM conditions enable the use of precision in policies and enhance users' understanding of these policies, creating permission management for the systems and thus bringing about more safety measures.

- **Compliance:** The creation of IAM conditions gives the companies an easier way to comply with data regulations and safety measures. Therefore, the restrictions ensure device compliance, time, and location properties match the desired levels.

- **Enhanced security:** A context-aware approach ensures companies have minimal intrusions and restricts access even among employees of the company, leading to reduced attack vectors on their devices (Veeramachaneni, 2025).

- **Operational flexibility:** Companies can adapt to changing conditions, instead of traditional approaches that affect the ways to identify and manage IAM policies.

Best Practices in IAM Conditions

IAM conditions have to be set to meet a company's needs. The following approaches have to be used:

- **Policy of least privilege:** This policy has to be implemented so minimum-level access is given to users to handle their duties effectively.

- **Regular audits:** Having a regular audit policy aligns IAM condition policies with current standards and requirements that match the organizational needs at all times. These conditions help carry out IAM management at a broader level.

- **Test before deployment:** IAM conditions have to be validated to confirm they work address the needed boundaries before they are scaled organization-wide.

- **Combined attributes:** Several attributes can be used to ensure increased access control. These aspects will ensure that there are different ways to engage users and verify requests (Talluri et al., 2023).

- **Monitor logs:** Access logs have to be monitored and analyzed to enhance the understanding of IAM conditions, leading to the identification of key areas of improvement.

- **User awareness:** To achieve stellar service, organizations have to educate users on IAM and work within the provided boundaries, ensuring sustainable results. Thus, user awareness is a key factor that will help in making advancements in critical resource areas.

IAM conditions are a great addition to organizations as they create context-aware access controls. These access controls increase security, making it easier to manage aspects such as location, device compliance, and time. Each of these factors creates helps handle device and data safety. IAM conditions have to mirror the access policies within organizations to enable the correct device management and data safety from users through access control.

Cross-Project and Cross-Organization Access with IAM Policies

Contemporary cloud environments enable organizations to carry out multiple projects and even collaborate in handling projects across different boundaries. Notably, handling access helps achieve goals set within these organizations. IAM ensures that the framework is well handled, leading to project and organizational management, reiterating the need to have higher possibility of control, operational efficiency and enhanced security in addressing the needed imperatives (Belchoir et al., 2020). Therefore, having Cross-project and cross-organizational access control can help companies to mitigate measures meant for high level of success as needed within the institutions.

Cross-Project and Cross-Organization Access

IAM policies help control access to organizational resources. They allow for detailed permissions based on user attributes and roles, enabling users to access what they need. Using the right IAM strategies ensures relevant and secure access across different projects and teams, promoting collaboration on cloud platforms. This requires careful management of key elements to ensure effective and secure access. These are the key elements:

- **Projects and organizations:** Projects are resource containers that hold services and containers. Projects can include aspects such as storage buckets. On the other hand, organizations are containers that can be used to group different projects into a common domain and help address their functionalities across various segments, as depicted in Figure 3-3.

CHAPTER 3 ADVANCED IAM FEATURES

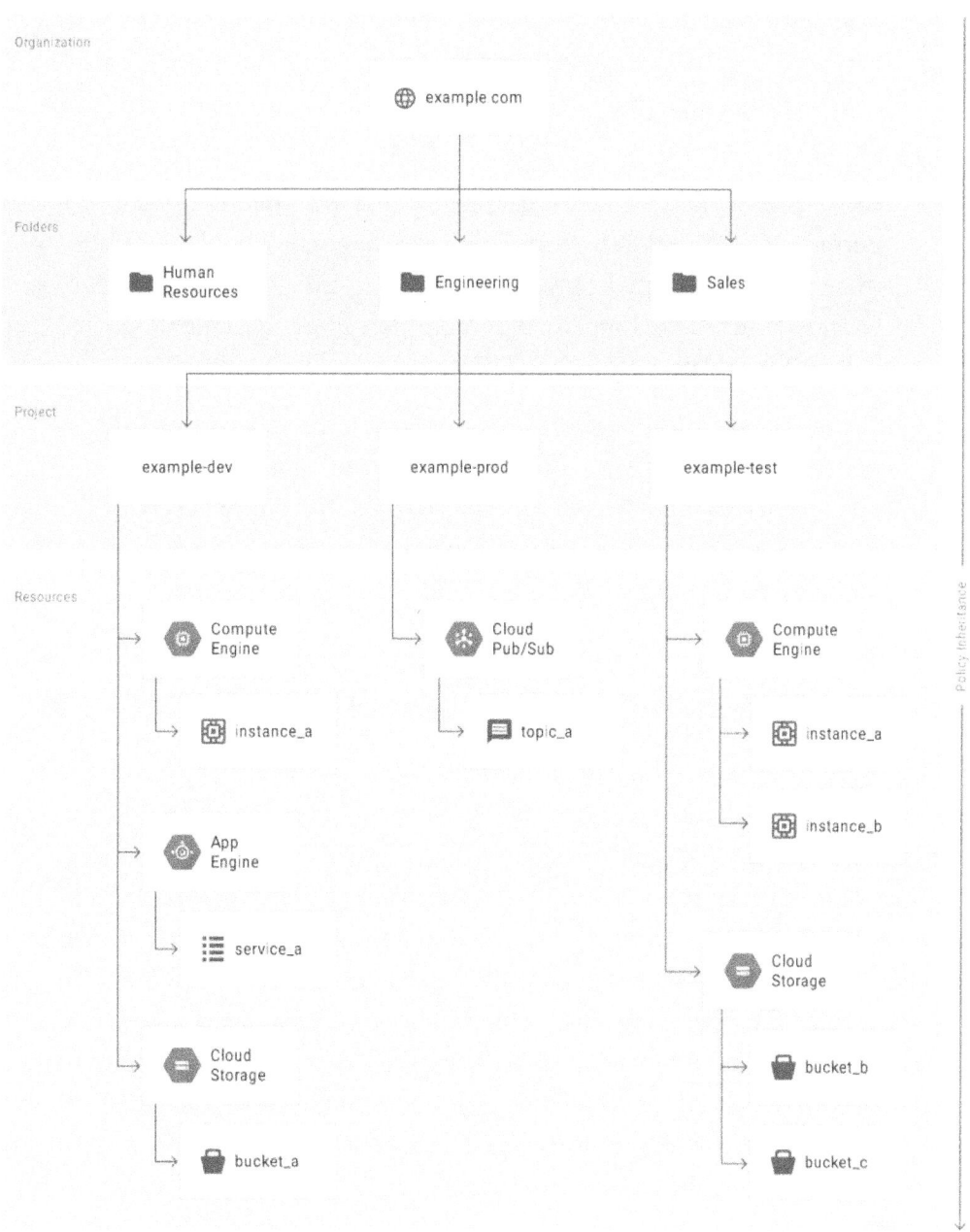

Figure 3-3. *Cross-project access on GCP*

- **IAM roles and permissions:** These are policies that include a set of permissions meant to enable users to address actions on resources. They can have a predefined capacity to handle the user requirements based on the cloud provider. In some cases, they have to be custom built to ensure critical engagement and the right categorization of information.

- **Cross-project access:** This is a type of access that enables users to perform tasks on other projects when they have credentials, allowing them to get into a single project. The identification is an instrumental way to figure out the functionality of entities, creating a reliable way to handle projects, as shown in Figure 3-3.

- **Cross-organization access:** Cross-organization access means IAM policies can ensure that boundaries for users are maintained to ensure collaboration between various parties. The approach creates a reliable way to manage information and grow with the increasing amount of information (Zhu et al., 2022).

Enabling Cross-Project and Cross-Organization Access

Cross-project and cross-organization access are critical to ensuring collaboration, so it is imperative for an organization to learn the right way to model the access. It is important for them to work together.

resources to complete their tasks. To support this, resources must be identified, planned for, and managed properly. The first step is to identify the resources needed, which helps determine what should be shared across projects (Herzig, 2020). The second step is to assign roles to service accounts that will request the resource. The creation of a service account in the project will help with access, ensuring that the resource will be accessed by both parties. The role will be assigned in the IAM policy within the resource, leading to the successful management of the entire system.

Cross-organization access can be conducted in various ways, as it is more complex and requires granting permission to groups, users, and service accounts in a different organization. This process demands precision to enable the collaboration between the organizations. Also, their IAM policies must be aligned. Thus, the cross-organization approach is an increasingly effective way to handle collaboration. The first step is to create a trust relationship between the organizations. The trust relationship makes it easier to ensure that the organizations can access each other's resources, as shown in Figure 3-4.

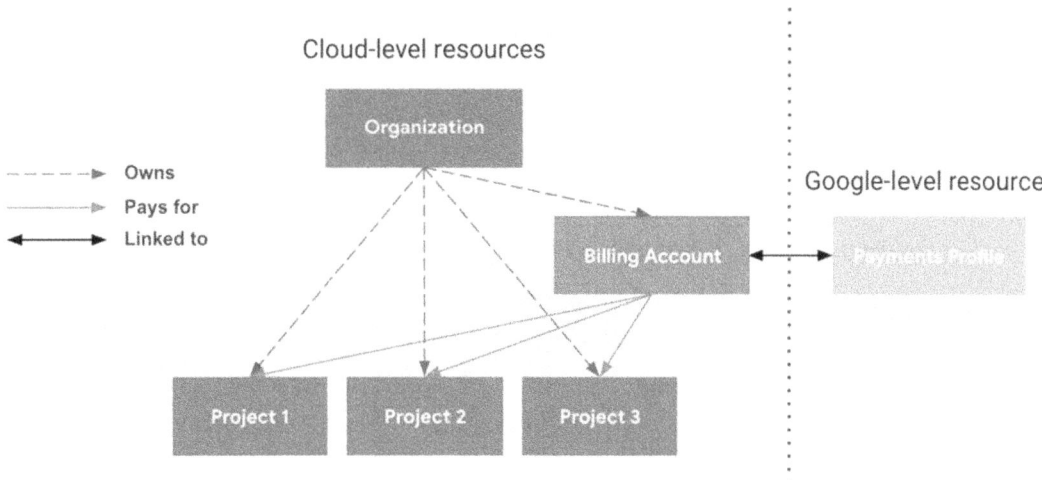

Figure 3-4. *Trust relationship and interaction on cross-organization access*

The use of domain or email identity will ensure that there is a way to handle information management between the organizations. This step grants access to a user's email address or the entire organization's domain. In addition, policies have to be assigned at the resource level to in the name of the least privilege principle. Continually assessing the privileges and auditing the logs creates better understanding and enhances the trust relationship.

Benefits of Enabling Cross-Project and Cross-Organization Access

Both cross-project and cross-organization access offer benefits:

- **Enhanced collaboration:** These approaches ensure that there is seamless handling and sharing of resources and data between partner organizations or even teams within the same company.

- **Scalability:** Each process ensures that IAM policies can accommodate companies with higher levels of collaboration. This implies that large organizations with several teams and projects can handle their activities with ease by using IAM policies to fit their structure (Zhang et al., 2019).

- **Centralized governance:** The use of IAM policies for collaboration ensures a central point of managing access control. This ensures that there is consistency in advancing security controls and meeting critical needs in whatever capacity has been offered.

- **Security:** Using either approach ensures a granulated access model, where roles are assigned and permissions are granted to these roles. These processes handle specific resources for specific people. Therefore, the security, with more audits, enables the organization to have an increased level of facilitating and advancing the right policies when needed.

- **Flexibility:** Several use cases can be applied, ranging from temporary project collaborations to long-term partnerships.

Best Practices for Cross-Project and Cross-Organization Access

Cross-project and cross-organization access need to be managed carefully to ensure proper identity verification and control. Using these techniques within an organization requires following proven practices to ensure successful outcomes. Some main approaches are as follows:

- **Least privilege:** This policy will ensure roles have the minimum permissions necessary for their assigned tasks. This will ensure that there is an increased level of safety in the organization and projects.

- **Service account automation:** Automated processes have to use service accounts. This will result in better management than when using individual user accounts. The approach lessens the risk of unauthorized data access when credentials have been leaked. Therefore, the approach increases security (Caballero & Piattini, 2022).

- **Regular audits:** Access policies have to be subjected to regular audits, enabling compliance to IAM policies and organizational security settings. This method creates the right scope and capacity to advance model development at the right scope and level for managing the organizational needs.

- **Conditional access policies:** Having conditional access policies creates the best way to define controls and restrict access in certain instances. This approach creates an instructional path to deal with the security and safety of projects, leading to fewer intrusion risks and attack vectors.

- **Monitor and log access:** Monitoring IAM-related activities will help detect and prevent unauthorized access. Effective access management allows for the early identification and resolution of issues, improving overall security and success in addressing potential problems.

- **Access review procedures:** Having the right access review procedures will help when validating ongoing activities for collaborators. Advanced monitoring can help an organization succeed.

Cross-project and Cross-Organization access with IAM policies is a key way to enhance collaboration and maintain security. Better governance through centralized operations on the cloud operations will ensure that organizations can achieve their intended goals easily. Choosing the best framework for sharing resources in projects and organizations further demands the use of robust monitoring, regular audits, and conditional policies. Implementing least privilege and compliance regulations will enhance collaboration. Therefore, these attributes will likely lead to a more secure and productive collaboration between projects and organizations.

Implementing Temporary and Time-Bound Access Securely

The contemporary cloud environment allows for time-bound access as well as temporary access to resources. This approach ensures that audits, operational tasks, collaboration with eternal entities, and troubleshooting can be conducted to assist in handling every integration correctly. Additionally, temporary access helps reinforce agility but also has security risks when not handled in an appropriate manner (Singh et al., 2023). Therefore, the improper handling of temporary access can lead to a data breach or unauthorized resource access. Therefore, organizations have to use secure and properly defined approaches to handle time-bound access.

Demand for Temporary and Time-Bound Access

Temporary and time-bound access have been used more often recently because of all the third-party collaborations within companies. There are consulting partners, contractors, and external vendors who need to access specific resources. And in some cases, emergency operations demand incident response teams to get immediate access to critical systems in an organization, leading to the increased use of time-bound access for these specific activities. Nonetheless, audits and compliance regulations mean that organizations have to give temporary access to external auditors and regulators (Cameron & Williamson, 2020). Finally, companies have to work with IT personnel who require elevated privileges when they are diagnosing, resolving, and maintaining the entire system. In these situations, time-bound access can assist.

Strategies for Secure Time-Bound Access

Time-bound access in organizations works within a framework that adjusts to the organization's needs. Secure handling of these needs is essential for effectively managing valuable system interactions. Considerably, policies that can be used include the following:

- **Time-bound IAM policies:** Enforcing time-bound policies will ensure that permissions expire automatically. After a specified duration elapses, the permissions turn on, ensuring that there is no access after that point. IAM policies can be configured on GCP to ensure that the duration is clearly defined and activities are conducted to achieve the desired value (Glocker et al., 2024).

- **Just-in-time access:** JIT access ensures that there is minimal risk within the system. It offers permissions only when requested, and permissions work only after specific approval. The JIT model ensures that access is revoked by the end of a predefined timeline or immediately after the task is completed, as indicated in Figure 3-5. JIT is therefore key to ensuring that the credentials are created and that they ask for permission that is then granted.

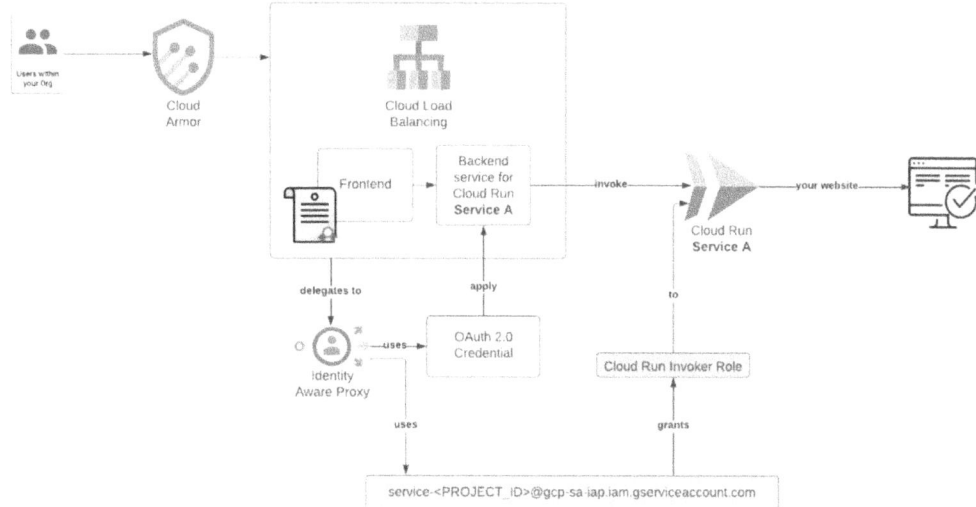

Figure 3-5. GCP JIT access

- **Leverage temporary credentials:** Using Google Cloud short-lived service accounts helps create temporary credentials. This features allows you to set credentials to expire automatically and reduces the risk of misuse.

- **Multifactor authentication (MFA):** MFA helps to validate users when they want to access sensitive data. Sensitive resources will be provided only to key users, ensuring that the framework can offer stellar service to the temporary users. MFA therefore adds a layer of protection.

Having temporary and time-bound access reduces the risk of prolonged access and over-privileged users. Having time-bound IAM policies, JIT access, and temporary credentials ensures that access can be provided when necessary and revoked after being used. Thus, this feature is the best way to handle temporary users. Table 3-1 summarizes the different practices that can help manage temporary access, where you want a balance between flexibility and security.

Table 3-1. Summary of Temporary and Time-Bound Access Types

Type	Implementation	Security Measures
Time-bound policies	Using IAM to have automatic timestamp	Automated expiration Audits
Temporary credentials	Short-lived tokens	Automated revocation
JIT access	Accepting requests as they are sent	MFA approval

Industry-Specific Examples of Context-Aware Access Policies in Healthcare and Education

Context-aware policies have been in use for several years. They have been used in different industries like healthcare and education to help manage location, role, device security, and time of access. They ensure that there is operational efficiency while dealing with various advances in the modern organizations.

Context-Aware Access in Healthcare

The healthcare industry has complex data management practices because of the sensitive nature of personal data in the sector. Data from clinical trials, patient records, and insurance details has to be protected in the most appropriate manner. Therefore, using context-aware policies to ensure healthcare environments have strict policies and specific steps to handle access is required. The use of context-aware access is conducted in the following situations:

- **Electronic health records:** EHR systems have sensitive information that has to be guarded. Context-aware access is implemented to ensure that a user's roles define the level of access to data that they have. For instance, this enables doctors to have access to patient medical history, nurses to have limited access to see vital statistics and care plans, and administrative staff to look at billing and insurance information. This approach ensures that role-based attributes are used in the context-aware approach in the medical sector (Thomas, 2022).

- **Telemedicine:** Telemedicine enables healthcare practitioners to access patient data. This works based on location settings, ensuring that the user is within the trusted location.

- **Mobile access:** Healthcare networks have to use device compliance to ensure that only specific devices are accessing networks. The approach works by setting the compliant devices to the network. Then noncompliant devices are denied from accessing the network.

Context-Aware Access in Education

Education has different stakeholders and uses student data to help ensure seamless service provision. A context-aware system can assist learners, administrators, and teachers. The context-aware approach works in the form of the following:

- **Time-based access to learning platforms:** Online learning uses time-based access to ensure that there is restricted access to examination content, before examinations take place. The time window is also restricted, ensuring that once it elapses, there is no more access to the platform (Josang, 2024).

- **Administrative portals:** Learning institutions can ensure that administrative portals are accessible only on campus network, ensuring that there are no external IP addresses that sign onto the network at any given point. This limits exposure to sensitive content and makes thee attack surface smaller.

Sensitive data requires advanced IAM features to ensure secure access. Using context-aware access policies allows seamless integration of information and addresses key variables. These policies consider factors like time, role, device compliance, and location to manage access effectively. As a result, context-aware policies build trust by protecting sensitive data while supporting efficient operations within the organization.

References

Arfaoui, A., Kribeche, A., & Senouci, S. M. (2019). Context-aware anonymous authentication protocols in the internet of things dedicated to e-health applications. *Computer Networks*, *159*, 23-36.

Belchior, R., Putz, B., Pernul, G., Correia, M., Vasconcelos, A., & Guerreiro, S. (2020, December). SSIBAC: self-sovereign identity based access control. In *2020 IEEE 19th International Conference on Trust, Security and Privacy in Computing and Communications (TrustCom)* (pp. 1935-1943). IEEE.

Caballero, I., & Piattini, M. (Eds.). (2023). *Data Governance: From the Fundamentals to Real Cases*. Springer Nature.

Cameron, A., & Williamson, G. (2020). Introduction to IAM Architecture (v2). *IDPro Body of Knowledge*, *1*(6).

Glöckler, J., Sedlmeir, J., Frank, M., & Fridgen, G. (2024). A systematic review of identity and access management requirements in enterprises and potential contributions of self-sovereign identity. *Business & Information Systems Engineering*, *66*(4), 421-440.

Herzig, T. W. (2020). Identity and Access Management. In *Information Security in Healthcare* (pp. 55-74). HIMSS Publishing.

Jøsang, A. (2024). IAM—Identity and Access Management. In *Cybersecurity: Technology and Governance* (pp. 191-214). Cham: Springer Nature Switzerland.

Qiu, J., Tian, Z., Du, C., Zuo, Q., Su, S., & Fang, B. (2020). A survey on access control in the age of internet of things. *IEEE Internet of Things Journal*, *7*(6), 4682-4696.

Singh, C., Thakkar, R., & Warraich, J. (2023). IAM identity Access Management—importance in maintaining security systems within organizations. *European Journal of Engineering and Technology Research*, *8*(4), 30-38.

Talluri, S., Anne, V. P., & Chadalavada, V. S. (2023). ROLE-BASED ACCESS CONTROL (RBAC) IN A CENTRALIZED IDENTITY AND ACCESS MANAGEMENT (IAM) SYSTEM. *International Journal of Information Technology (IJIT)*, *4*(1).

Thomas, H. (2022). IAM and AI: A Dual Approach to Securing Healthcare Cloud Infrastructures for HIPAA Compliance.

Veeramachaneni, V. (2025). Integrating Zero Trust Principles into IAM for Enhanced Cloud Security. *Recent Trends in Cloud Computing and Web Engineering*, *7*(1), 78-92.

Xiao, S., Ye, Y., Kanwal, N., Newe, T., & Lee, B. (2022). SoK: context and risk aware access control for zero trust systems. *Security and Communication Networks*, *2022*(1), 7026779.

Zhang, Y., Krishnan, R., Patwa, F., & Sandhu, R. (2019). Access Control in Cloud IaaS. *Security, Privacy, and Digital Forensics in the Cloud*, 81-108.

Zheng, R., Jiang, J., Hao, X., Ren, W., Xiong, F., & Zhu, T. (2019). CaACBIM: A context-aware access control model for BIM. *Information*, *10*(2), 47.

Zhu, Y., Wu, X., & Hu, Z. (2022). Fine grained access control based on smart contract for edge computing. *Electronics*, *11*(1), 167.

CHAPTER 4

Service Accounts and Workload Identity

I. Service Accounts: Types and Best Practices

Service accounts are key to the performance of Identity And Access Management (IAM) in GCP. Service accounts help virtual machines, applications, and nonhuman entities interact with GCP services while maintaining efficiency and security. Service accounts have automated workloads and processes, unlike user accounts, which are obviously used by individuals. Service accounts in GCP can be used for different purposes to help with security, management, and scalability of the platform.

Types of Service Accounts

There are different types of service accounts:

- **Default service accounts:**

    These are accounts that are automatically created on GCP when different services are enabled. Services that create these default accounts include App Engine, Compute Engine, and Cloud Functions (Wang, 2024). The accounts have predefined permissions and roles that enable them to function correctly and meet the needs of the organization. The service accounts have different features such as the following:

    - **Broad permissions:** The default service accounts have broad permissions so they can interact with different services.

CHAPTER 4 SERVICE ACCOUNTS AND WORKLOAD IDENTITY

- **Automatic creation:** GCP automatically creates these service accounts whenever a service has been enabled on the platform.

- **Limited customization:** The default service accounts have a limited ability to modify roles; they work only with the predefined functionality.

Service accounts are used in different instances. They can be used to enable managed services and provide much of the underlying infrastructure. Service accounts have a quick setup and can run without extensive configurations to meet their needs, as indicated in Figure 4-1.

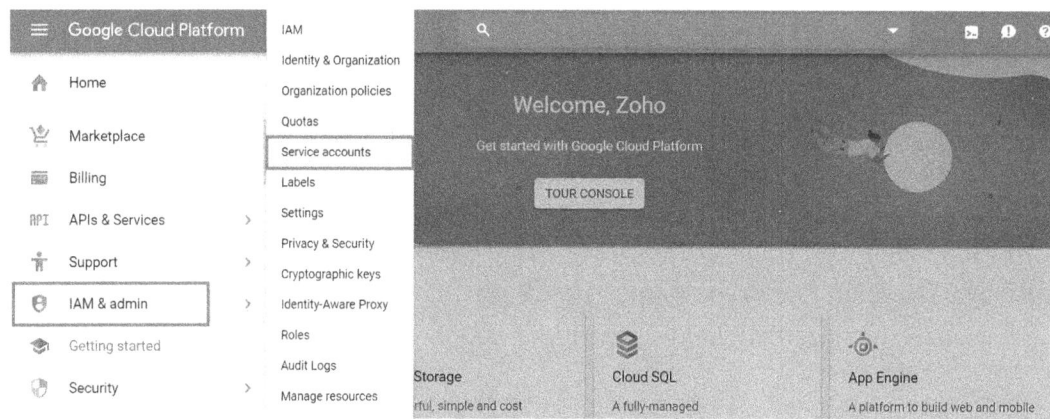

Figure 4-1. *GCP default service accounts*

- **User-managed service accounts:**

    GCP users can create and manage service accounts, providing specific permissions to virtual machines, applications, and workloads. The accounts provide more flexibility and control than the default service accounts. They have different features such as the following:

CHAPTER 4   SERVICE ACCOUNTS AND WORKLOAD IDENTITY

- **Granular control:** User-managed service accounts enable the provision of fine-grained access control, ensuring that the principle of least privilege is being followed.

- **Manual creation:** User-managed service accounts have to be manually created, offering a wide range of decisions for the user to design controls and other needs. They can be created with the CLI, GCP Console, and APIs.

- **Custom roles and permissions:** User-managed service accounts can have custom roles and permissions, enabling them to meet the specific needs of the provided workload.

User-managed service accounts can be used in multitenant applications, which ensures that the different workloads have distinct levels of access control (Kingsley, 2023). Additionally, they can be used in production workloads and provide security and compliance. To create these service accounts, you can use the following code:

```python
from google.oauth2 import service_account
from googleapiclient.discovery import build

# Replace with your project ID and service account details
PROJECT_ID = "your-project-id"
SERVICE_ACCOUNT_NAME = "new-service-account"
DISPLAY_NAME = "New Service Account"
ROLE = "roles/storage.objectViewer"

# Authenticate using a service account key
credentials = service_account.Credentials.from_service_account_file("path/to/service-account-key.json")
service = build("iam", "v1", credentials=credentials)

# Create a new service account
def create_service_account():
    service_account_body = {
        "accountId": SERVICE_ACCOUNT_NAME,
        "serviceAccount": {
```

CHAPTER 4   SERVICE ACCOUNTS AND WORKLOAD IDENTITY

```python
            "displayName": DISPLAY_NAME
        }
    }
    request = service.projects().serviceAccounts().create(
        name=f"projects/{PROJECT_ID}",
        body=service_account_body
    )
    response = request.execute()
    print("Service Account Created:", response)

# Assign a role to the service account
def assign_role_to_service_account():
    policy = service.projects().getIamPolicy(resource=PROJECT_ID).execute()
    binding = {
        "role": ROLE,
        "members": [f"serviceAccount:{SERVICE_ACCOUNT_NAME}@{PROJECT_ID}.
        iam.gserviceaccount.com"]
    }
    policy["bindings"].append(binding)
    request = service.projects().setIamPolicy(resource=PROJECT_ID,
    body={"policy": policy})
    request.execute()
    print(f"Role {ROLE} assigned to {SERVICE_ACCOUNT_NAME}")

# List all service accounts in the project
def list_service_accounts():
    request = service.projects().serviceAccounts().list(name=f"projects/
    {PROJECT_ID}")
    response = request.execute()
    for account in response.get("accounts", []):
        print(account["email"])

# Run the functions
create_service_account()
assign_role_to_service_account()
list_service_accounts()
```

- **Google-managed service accounts:**

    Google-managed service accounts are internal and are used by Google to enable the management and operation of GCP services. The accounts cannot be edited or configured by users, and they have the following features:

    - **No user interaction:** Google-managed service accounts do not appear within IAM policy bindings, and users cannot interact with the accounts directly.

    - **Internal use:** The accounts are used by Google to assist in service management. Services that can be managed include Cloud Storage and BigQuery.

    - **High security:** Google-managed service accounts are very secure and can make critical operations within GCP easy to interact with.

    Service accounts can be enable managed services to operate smoothly and securely. Service accounts can also be used for internal operations such as backup, maintenance tasks, and data replication activities.

Best Practices in Service Accounts

There are different strategies to make the most of service accounts. Key strategies for using service accounts include the following:

1. **Principle of least privilege:**

    This is a security practice related to providing only the minimum needed permissions to ensure a service account handles its tasks. Minimizing risk in this way makes it harder for accidental or malicious actions that affect the entire GCP environment. The implementation of least privilege stems from having custom roles with only the necessary permissions. Having regular audits ensures that the permissions align to users' activities within the system (Sukhdeve D & Sukhdeve S, 2023).

CHAPTER 4 SERVICE ACCOUNTS AND WORKLOAD IDENTITY

2. **Separate service accounts for Different workloads:**

   Users have to configure distinct service accounts to be used with different environments, applications, and workloads. This not only improves management but enhances security. This can be handled by ensuring that there are environment and application service accounts. They assist in offering distinct capacities and levels of administering and ensuring service account use across all levels. In addition, workload-specific accounts have to be configured to ensure they have different batch jobs, data processing pipelines, and web servers to be applied as required.

3. **Auditing and monitoring of service account use:**

   Continuous monitoring and regular audits means you can detect and respond to malicious activities quickly. GCP supports several tools to this. For example, Cloud Audit Logs, Cloud Monitoring, and IAM Recommender can assist in managing and addressing the management needs of service accounts.

4. **Avoidance of default service accounts:**

   These service accounts have broad permissions, which enable malicious individuals to attack platforms in different ways. Therefore, preventing these broad permissions will lead to better safety and security for the service accounts. This can be done by disabling default accounts, ensuring that there are custom accounts, and applying role assignments.

5. **Use of short-lived credentials and having workload identity federation:**

   Users have to use workload identity federation and short-lived credentials to reduce security risks. Enabling short-lived tokens ensures there are no long-living keys, as indicated in Figure 4-2. Additionally, practicing key rotation allow you to rotate service account keys, monitoring service accounts and preventing any breaches that might affect the system.

CHAPTER 4 SERVICE ACCOUNTS AND WORKLOAD IDENTITY

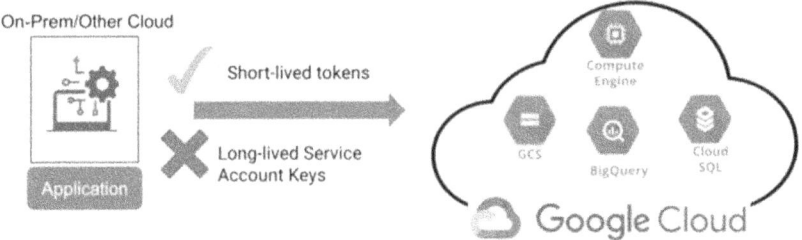

Figure 4-2. Short-lived credentials and WIF on GCP

6. **Secure service account keys:**

    Service account keys have to securely stored with restricted access. A secret manager can store the service account keys, ensuring that there is limited access to them. Practicing key rotation and access restrictions are also important.

7. **Implementing service account impersonation:**

    Service account impersonation ensures that a single service account can assume the roles and permissions of another. This allows you to delegate access to other service accounts without having to share the keys between them. You can grant temporary access for specific tasks and then revoke it.

Service accounts can improve the security and efficiency of transactions between cloud resources and workloads. Following best practices with service accounts leads to better management, scalability, and security of the GCP environment (Roy et al., 2021). These approaches help delegate service account responsibilities and ensure effective management to improve performance and track progress within the GCP environment.

II. Workload Identity Federation for Secure Service Access

Modern cloud environments have to work with multiple services on different platforms. The workloads need to be secure and easy to manage. GCP uses workload identity federation to ensure workloads running outside GCP can access services and resources without any long-lived service account keys. Workload identity federation ensures that

CHAPTER 4   SERVICE ACCOUNTS AND WORKLOAD IDENTITY

tools such as Azure, AWS, and Kubernetes clusters can access to the GCP resources. It also makes it easier to work with external identity providers (IdPs) to provide access to GCP resources (Spinella, 2025). the following are key concepts associated with workload identity federation:

a. **External identity providers:** These are systems that handle identities and provide the tokens needed for authentication. They include Azure Active Directory and AWS IAM.

b. **Workload identity pool:** This is a resource available on GCP that acts as a bridge between external IdPs and the GCP platform. The pool ensures that GCP can recognize and provide trust to tokens generated from the external IdPs.

c. **Token exchange:** This is a process that ensures an external IdP token is provided for a change with GCP token, ensuring that the platforms can work with GCP resources.

d. **Attribute exchange:** This approach provides attributes from external IdP tokens to the GCP identity mechanism, as depicted in Figure 4-3, ensuring that the right permissions are requested and granted as needed.

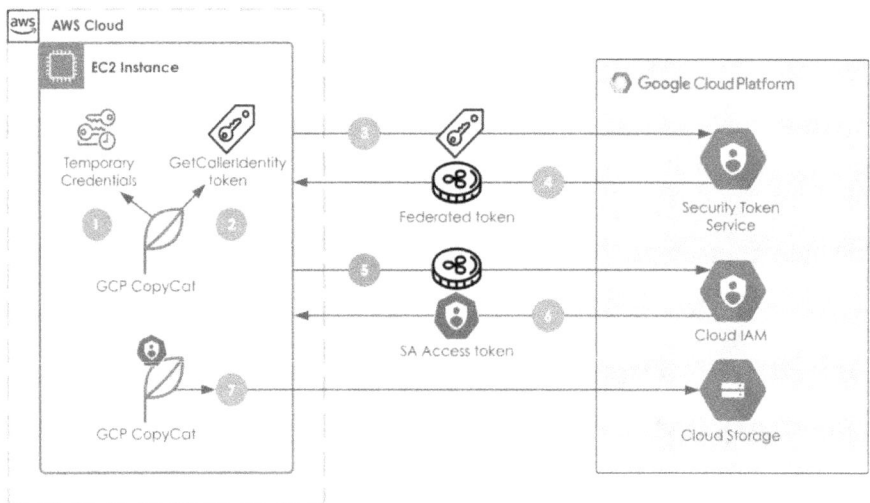

Figure 4-3. *Functionality of WIF on GCP and AWS*

Benefits of Workload Identity Federation

Workload identity federation has several benefits for service accounts. These benefits include as the following:

a. **Centralized identity management:** Enterprises can centralize identity management on GCP, making access control and security policies more consistent and easier to use.

b. **Elimination of need for long-lived keys:** This is an instrumental role where the WIF provides just the right amount of access, reducing the risk of exposure for the GCP platform (Somasundaram, 2024).

c. **Enhanced security:** WIF works to ensure that there are several security advances such as multifactor authentication and access control through conditional policies.

d. **Support for multicloud and hybrid environments:** WIF makes it seamless to manage activities on several different cloud platforms and on-premise environments.

e. **Simplified key management:** Service account key rotation is a must when working with WIF to ensure there are automated management approaches for short-lived tokens.

Enabling WIF

WIF can be enabled through a series of steps that ensure the configuration of external IdPs, mapping attributes, and workload identity pools (Salah et al., 2024). The key steps are as follows:

i. Configure an external identity provider.

ii. Create a workload identity pool.

iii. Map attributes from the external tokens in use.

iv. Set up a token exchange.

v. Access GCP resources.

CHAPTER 4   SERVICE ACCOUNTS AND WORKLOAD IDENTITY

To enable WIF, you can use the following Python code:

```python
from google.auth.transport.requests import Request
from google.oauth2 import id_token

# Replace with your Workload Identity Pool and Provider details
WORKLOAD_IDENTITY_POOL = "projects/your-project-id/locations/global/workloadIdentityPools/your-pool-id"
SERVICE_ACCOUNT_EMAIL = "your-service-account@your-project-id.iam.gserviceaccount.com"

# Fetch an ID token from the external identity provider
def fetch_external_token():
    # Replace with your external token fetching logic
    return "your-external-id-token"

# Exchange the external token for a GCP access token
def exchange_token(external_token):
    request = Request()
    token = id_token.fetch_id_token(request, WORKLOAD_IDENTITY_POOL, external_token)
    print("GCP Access Token:", token)

# Main function
def main():
    external_token = fetch_external_token()
    exchange_token(external_token)

if __name__ == "__main__":
    main()
```

Real-World Use

WIF is used in various scenarios to meet different needs, allowing for control, customization, and risk mitigation of specific functions. Examples include the following:

- a. **Hybrid cloud environments:** WIF can be applied within a hybrid setting to work on both on-premise settings and in-cloud instances. WIF can ensure that there is a seamless access to GCP resources.

b. **Multicloud workloads:** Organizations can work with several cloud platforms, and WIF allows them to use all cloud platforms on a single credential.

c. **CI/CD pipelines:** CI/CD makes it easier to handle GCP resources, deploy applications, and meet the infrastructure management needs. Using WIF ensures that the CI/CD pipelines have access to GCP resources through credentials made available to platform.

d. **Kubernetes workloads:** Workloads on the Kubernetes clusters can use WIF to make sure there is access to every GCP resource. The service account keys can be stored within the Kubernetes secrets, further lowering threats of exposure.

Implementing Workload Identity Federation

There are many ways to improve how WIF functions in an enterprise. You want to ensure the WIF platform delivers effective results, enabling better use of WIF across any cloud platform, by following these best practices:

a. **Regular audits of workload identity pools:** Conduct regular audits to make sure the configuration is correct and the GCP resources can be accessed by the authorized workloads.

b. **Short-lived tokens:** External IdPs use short-lived tokens to reduce any imminent risks of misuse of privilege and ensure increased access for the required duration.

c. **Attribute-based access control:** This ensures fine-grained access control so workloads only have the permissions designated to them to handle the required work in the cloud platform.

d. **Secure external IdPs:** External IdPs have to be secured well. Conditional access and MFA should be used (Dasher et al., 2022).

e. **Monitor token use:** GCP monitoring and logging tools can track token usage. This will help detect any unusual or unauthorized activities on the platform. The monitoring features will send alerts when seeing any suspicious behavior on the platform.

f. **Document configuration and process needs:** WIF has to be configured in a manner that helps with token exchange, attribute mapping, and access control policies. This provides consistency and makes troubleshooting easier.

g. **Testing in nonproduction environments:** WIF can provide a nonproduction environment. Any misconfigurations experienced in the testing environment can be minimized, ensuring the best result in production.

WIF provides security and efficiency when providing GCP resources to external service providers. GCP resources help manage service account keys, integrate external identity providers, and simplify operations in multicloud and hybrid cloud environments. These benefits enable WIF to deliver the right outcomes when working with service accounts. Thus, by following best practices, WIF enhances the functionality of cloud platforms.

III. Managing and Rotating Service Account Keys

Service account keys are important when using Identity and Access Management (IAM) in GCP. They help applications interact with GCP services and workloads with authentication and access management (Mulder, 2020). Service account keys, however, are security risks because they are long-lived credentials on systems, making them vulnerable to unauthorized access, security incidents, and data breaches.

Service account keys have several benefits when working with the GCP platform. These keys, however, have the challenge of broad permissions, making them hard to navigate and work with. In addition, service accounts are difficult to monitor. The following are strategies to manage service account keys:

a. **Secure key storage strategies:** Service account keys have to be securely stored to work across multiple cloud platforms. Storage can be secured through encryption and Google Cloud Secrets Manager.

b. **Minimal service account key use:** Using service account keys will ensure you have more protection. Different authentication models can be used to secure service accounts keys. Using impersonation, short-lived credentials and WIF create the best chance to secure and manage service account keys.

CHAPTER 4   SERVICE ACCOUNTS AND WORKLOAD IDENTITY

c. **Monitoring key use:** Continuous monitoring enables you to detect anomalies early and can help with reporting. Different platforms can be used for monitoring, such as Cloud Audit Logs, IAM Recommender, and Cloud Monitoring (Cabianca, 2024).

d. **Restricting access to keys:** Having limited access to the service account keys ensures that only those who need them can access and use them. IAM policies can enforce access controls and provide avenues for further protection. Enterprises can use RBAC, the least privilege principle, and auditing of logs to assist in fine-tuning the access.

e. **Key rotation activities:** Key rotation can decrease the risk of compromise by never allowing any compromised keys to be used on the platform again. Strategies include automated rotation, grace periods for expiry, and manual rotation.

Service Account Key Rotation

The process of rotation involves creating a new key, updating the application to use the new key, and deleting the old key. The entire process needs to be secure. You can generate a new key in the IAM and Admin section in GCP, selecting the service accounts and selecting a specific account for which the key is rotated to. After selecting the service account, click the Keys tab to generate a new key and then select the type of key to generate. After this step, the application is updated by modifying configuration files, environment variables, and Kubernetes secrets to manage the new key. The old key is then deleted (Song, 2023). Constant monitoring and auditing of the keys should be done. You can automate the service account key rotation by using the following Python code:

```python
from google.oauth2 import service_account
from googleapiclient.discovery import build

# Replace with your project ID and service account email
PROJECT_ID = "your-project-id"
SERVICE_ACCOUNT_EMAIL = "your-service-account@your-project-id.iam.gserviceaccount.com"
```

```python
# Authenticate using a service account key
credentials = service_account.Credentials.from_service_account_file("path/
to/service-account-key.json")
service = build("iam", "v1", credentials=credentials)

# Generate a new service account key
def generate_new_key():
    request = service.projects().serviceAccounts().keys().create(
        name=f"projects/{PROJECT_ID}/serviceAccounts/{SERVICE_ACCOUNT_
        EMAIL}",
        body={}
    )
    response = request.execute()
    print("New Key Generated:", response["privateKeyData"])

# List all keys for the service account
def list_keys():
    request = service.projects().serviceAccounts().keys().list(
        name=f"projects/{PROJECT_ID}/serviceAccounts/{SERVICE_
        ACCOUNT_EMAIL}"
    )
    response = request.execute()
    for key in response.get("keys", []):
        print("Key ID:", key["name"])

# Delete a specific key
def delete_key(key_id):
    request = service.projects().serviceAccounts().keys().
    delete(name=key_id)
    request.execute()
    print(f"Key {key_id} deleted")

# Rotate keys
def rotate_keys():
    # Generate a new key
    generate_new_key()
```

```
# List all keys
list_keys()

# Delete the old key (replace with the actual key ID)
old_key_id = "projects/your-project-id/serviceAccounts/
your-service-account/keys/old-key-id"
delete_key(old_key_id)
```
```
# Run the rotation
rotate_keys()
```

Strategies for Key Management

Key management can be done in different ways to maximize benefits. The following approaches help organizations build sustainable and reliable systems, delivering effective results across various use cases:

A. **Use of key management service:** KMS can be used on GCP to encrypt and manage service account keys. This provides extra protection even when carrying out key rotation. It provides centralized management, ensuring encryption and auditing are being conducted (Shashi, 2023), as shown in Figure 4-4.

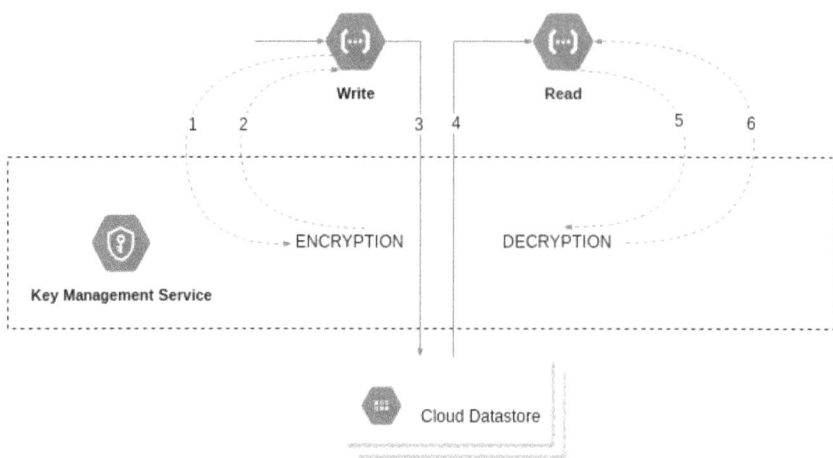

Figure 4-4. Using KMS to handle service account keys

> B. **Automated key rotation:** Automating key rotation means there is less chance of human error. Automation can be set up through custom scripts that assist in generating, deploying, and deleting keys in a consistent manner. Terraform can also be used to automate key creation and rotation. Integrating key rotation on the CI/CD pipeline is also key to the deployment process.
>
> C. **Conditional access policies:** Conditional access policies can help restrict key use. Conditions such as user roles, time of day, and IP ranges can be applied to further protect the service account keys. These regulations make keys more secure and ensure they cannot be misused if they are stolen.

Rotating service account keys is important when securing the GCP environment. These keys have to be used to access various services. Minimal use of long-lived keys, secure storage, and a restricted environment all help to secure the service account keys (Ipsale & Gilioli, 2022). Strategies such as automation, conditional access, and KMS will provide the service account keys with much needed security. But, the security mechanisms have to be applied to achieve the best results.

References

Wang, I. (2024). Provisioning Infrastructure on GCP. In *Terraform Made Easy: Provisioning, Managing and Automating Cloud Infrastructure with Terraform on Google Cloud* (pp. 95-170). Berkeley, CA: Apress.

Kingsley, M. S. (2023). Google Cloud Platform (GCP) Lab. In *Cloud Technologies and Services: Theoretical Concepts and Practical Applications* (pp. 325-378). Cham: Springer International Publishing.

Sukhdeve, D. S. R., & Sukhdeve, S. S. (2023). Introduction to GCP. In *Google Cloud Platform for Data Science: A Crash Course on Big Data, Machine Learning, and Data Analytics Services* (pp. 1-9). Berkeley, CA: Apress.

Roy, A., Banerjee, A., & Bhardwaj, N. (2021). A study on google cloud platform (gcp) and its security. *Machine Learning Techniques and Analytics for Cloud Security*, 313-338.

Spinella, E. F. (2025). Kubernetes Workload Identity Federation.

Somasundaram, P. (2024). Unified Secret Management Across Cloud Platforms: A Strategy for Secure Credential Storage and Access. *Int. J. Comput. Eng. Technol*, 15, 5-12.

Salah, M. A. B. H., Laborde, R., Benzekri, A., Kandi, M. A., & Ferreira, A. (2024). Identity management in cross-cloud environments: Toward self-sovereign identities using current solutions.

Dasher, G., Envid, I., & Calder, B. (2022). Architectures for Protecting Cloud Data Planes. *arXiv preprint arXiv:2201.13010*.

Mulder, J. (2020). *Multi-Cloud Architecture and Governance: Leverage Azure, AWS, GCP, and VMware vSphere to build effective multi-cloud solutions*. Packt Publishing Ltd.

Cabianca, D. (2024). Configuring Access. In *Google Cloud Platform (GCP) Professional Cloud Security Engineer Certification Companion: Learn and Apply Security Design Concepts to Ace the Exam* (pp. 15-175). Berkeley, CA: Apress.

Song, L. (2023). *The Self-Taught Cloud Computing Engineer: A comprehensive professional study guide to AWS, Azure, and GCP*. Packt Publishing Ltd.

Shashi, A. (2023). Advanced GCP Services. In *Designing Applications for Google Cloud Platform: Create and Deploy Applications Using Java* (pp. 155-180). Berkeley, CA: Apress.

Ipsale, M., & Gilioli, M. (2022). *Google Cloud Certified Professional Cloud Network Engineer Guide: Design, Implement, Manage, and Secure a Network Architecture in Google Cloud*. Packt Publishing Ltd.

CHAPTER 5

Securing API and Workloads

I. IAM Permissions for API Gateway, Cloud Functions, and Cloud Run

Contemporary cloud-native ecosystems require security for the APIs and workloads as well as confidentiality, integrity, and availability for the applications. Identity and Access Management (IAM) defines access levels for the various resources on GCP. Specifically, GCP services such as Cloud Functions, API Gateway, and Google Cloud use IAM to define security and manage the environment.

IAM Use in Google Cloud

IAM provides resource management on Google Cloud. IAM defines roles and permissions so that only specific users can perform specific actions on resources. GCP has the following key roles:

  a. **Predefined roles:** These are granular roles, which deal with specific services within the system such as role/cloudfunctions.developer.

  b. **Primitive roles:** These roles cover general requirements within the system. They are basic roles such as Editor, Owner, and Viewer, ensuring the incremental access and management of a control (Mohammed, 2019).

  c. **Custom roles:** These roles can be customized to meet particular requirements.

IAM roles can be provisioned to particular identities such as service accounts, groups, and users. The resource hierarchy includes the organization, projects, and folders.

IAM Permissions in Using API Gateway

API Gateway allows you to create, deploy, and manage APIs to work with backend services. IAM offers security to API Gateway deployments and configurations and prevents unauthorized access, as indicated in Figure 5-1. The IAM permissions are therefore key to ensuring that API Gateway is well secured and protected. Key IAM roles on the API gateway include the following:

- **roles/apigateway.admin:** This ensures that all API Gateway permissions are provided to the role. These permissions include the editor role and that one can manage IAM policies on these resources.

- **roles/apigateway.viewer:** This is a role that provides only read-only access to API Gateway configurations. This permissions enables the capacity to view API definitions, configurations, and gateways.

- **roles/apigateway.editor:** This offers permissions to create, update, and delete API Gateway configurations and deployments.

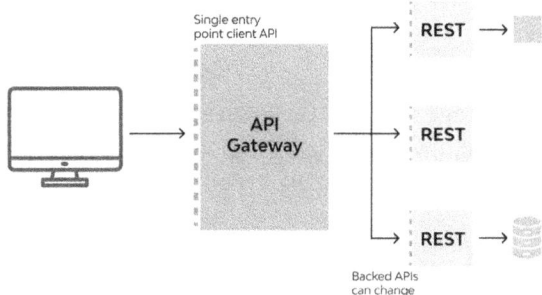

Figure 5-1. *API Gateway*

To work with IAM permissions on IAM, different practices can be following. First, following the principle of least privilege ensures that the minimum possible access is provided to users. Additionally, service accounts can be used to ensure backend integration activities (Kendyala, 2023). Integrating API Gateway with Cloud Run and

Cloud Functions is key to enable secure communication and limited permissions. Using Audit Logs allows you to monitor and change the API Gateway configuration based on imminent issues. These methods allow you to work with API Gateway through IAM configurations and get the best results.

IAM Permissions for Cloud Functions

Cloud Functions on GCP is a serverless environment meant to help with execution, enabling you to use code to respond to events. Cloud Functions has to be secured so that only allowed users can deploy, modify, and invoke functions on the channel. Different IAM roles can be used on Cloud Functions. the following are the main roles that can be applied:

- **roles/cloudfunctions.admin:** This role grants permission to manage IAM policies and functions.

- **roles/cloudfunctions.developer:** This provides permissions to create, delete, and update functions and manage the triggers and variables within the environment.

- **roles/cloudfunctions.viewer:** This offers read-only access to ensure users can view function metadata and configurations.

- **roles/cloudfunctions.invoker:** This enables the user to invoke functions; however, they lack the capacity to deploy and even modify the functions that they invoke.

When using IAM with Cloud Functions, different approaches are needed to effectively manage IAM requirements and support ongoing improvements. Using secure function triggers will ensure that only authorized users and services can trigger functions (Muppa, 2024). Additionally, the environment variables and secrets have to ensure that there is secure storage of sensitive information using Google Secret Manager so that the function's service account can access these secrets. In addition, having separate deployment and invocation roles helps keep things secure. The IAM permissions can be enforced using the following Python code:

```
import functions_framework
from google.auth.transport import requests
from google.oauth2 import id_token
```

```python
# The expected audience (client ID) for the token
AUDIENCE = 'your-client-id'

@functions_framework.http
def iam_secured_function(request):
    # Get the access token from the request headers
    auth_header = request.headers.get('Authorization')
    if not auth_header or not auth_header.startswith('Bearer '):
        return 'Unauthorized: Missing or invalid token', 401

    access_token = auth_header.split('Bearer ')[1]

    try:
        # Validate the access token
        token_info = id_token.verify_oauth2_token(access_token, requests.
        Request(), AUDIENCE)
        print("Token info:", token_info)

        # Check if the caller has the required role
        if 'roles/cloudfunctions.invoker' not in token_info.
        get('roles', []):
            return 'Forbidden: Insufficient permissions', 403

        # Proceed with the function logic
        return 'Hello, World! This function is secured with IAM.', 200
    except ValueError as e:
        return f'Unauthorized: {str(e)}', 401
```

IAM Permissions for Cloud Run

Cloud Run is for managing stateless containers. This managed compute platform ensures that stateless containers can function on GCP, allowing them to scale and integrate with other services (Roy et al., 2021). Cloud Run demands several roles in order to work. Thus, the permissions and roles that can be applied on Cloud Run include the following:

- **roles/run.admin:** This ensures that all permissions to help with the service management and presentation of IAM policies are available.

- **roles/run.invoker:** This role allows for the invocation of services without the capacity for deployment and modification.

- **roles/run.developer:** This offers permissions for the deployment and management of services that must perform traffic splitting and configuration updates.

- **roles/run.viewer:** This function gives read-only access to view the configurations and logs for various services.

Working with these IAM roles on the Cloud Run service is important. Using service accounts for containers is also key to ensure that a service account is assigned for every Cloud Run service. Monitoring and auditing Cloud Run services helps prevent unauthorized access, supports key functions, and ensures effective management of requirements (Borra, 2024). These sustainable models improve Cloud Run performance. IAM conditions can also be used to limit public access, enabling secure logins and better platform management.

Managing Service Accounts

When working with IAM on API Gateway, you can use Cloud Run, Cloud Functions, and service accounts to achieve the required results. Service accounts have to be designed precisely to ensure they abide by the principle of least privilege. This means they will have the least number of resources to achieve their duties at all time (Guptha & Murali, 2021). In addition, impersonation provides offers temporary access to resources to ensure that private keys are not shared but can still perform their duties, as indicated in Figure 5-2. Key management is another critical component, ensuring that the use of long-lived service account keys are not used. Instead, short-lived credentials and workload identity federation should be used.

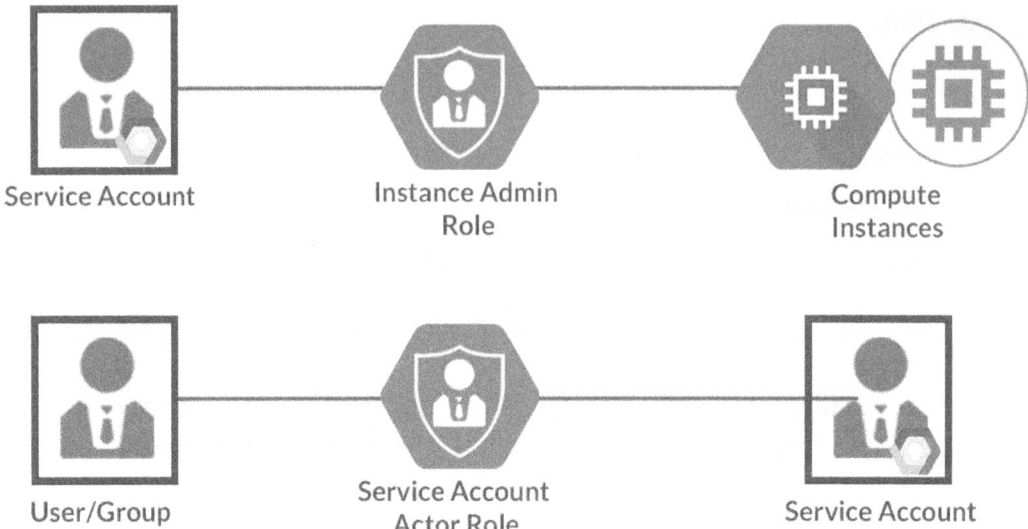

Figure 5-2. Google service accounts

Troubleshooting IAM Issues

IAM might experience issues that continually affect its functionality. To troubleshoot, the first step is to use IAM conditions to evaluate the problem. Also, auditing logs can help to determine root causes for major issues on the platform (Kukreti, 2023). Configuring IAM policies can be instrumental in correcting issues with the configuration of policies.

IAM permissions offer a lot of benefits to APIs and workloads on the GCP platform. The roles and practices on Cloud Functions, API Gateway, and Cloud Run provide critical access control mechanisms meant to ensure the secure management of resources. Of course, following best practices is always key in managing IAM permissions to achieve a secure, compliant, and beneficial IAM rollout on APIs and Workloads.

II. Service-to-service Authentication Using OAuth2.0 and API Keys

Service-to-service communication is key to building modular applications and enabling scalability within each of these applications. The use of services such as Cloud Run, API Gateway, and Cloud Function requires a system of authenticated and secure communication, which can be conducted only using API keys and OAuth 2.0.

Service-to-Service Authentication

Service-to-service authentication makes it so that only authorized services can communicate with one another. Service-to-service authentication is key in securing distributed systems and making sure no improper communication is allowed. It prevents data breaches and communication disruptions.

Service-to-service communication can be conducted using either OAuth 2.0 or API keys. OAuth 2.0 offers token-based authentication advances (Triartono & Negara, 2019). Authorization can also be used. Additionally, using API keys is a simple and less secure approach that identifies and authenticates services and applications.

Using OAuth 2.0 for Service-to-Service Authentication

OAuth 2.0 is an industry standard for authorization. It enables a client service to access resources instead of the resource server service. It does this without having to call for credentials from either of the services. Therefore, the access tokens are validated by OAuth 2.0, which provides permissions.

OAuth 2.0 works by allowing access tokens via the use of JSON Web Token (JWT), which contains information on the client, token expiration, time, and permissions granted to the client, as shown in Figure 5-3. The client credentials flow enables the access token from the authorization server based on the presentation of both the client secret and the client ID. This approach validates the credentials and provides an access token that can be used. OAuth 2.0 is beneficial for managing the authorization mechanism. Resource server validation ensures that the client has an access token within the HTTP authorization (Vasudevan, 2023). The resource server therefore provides validation for the token, granting access or even denying it after considering the permissions provided.

CHAPTER 5   SECURING API AND WORKLOADS

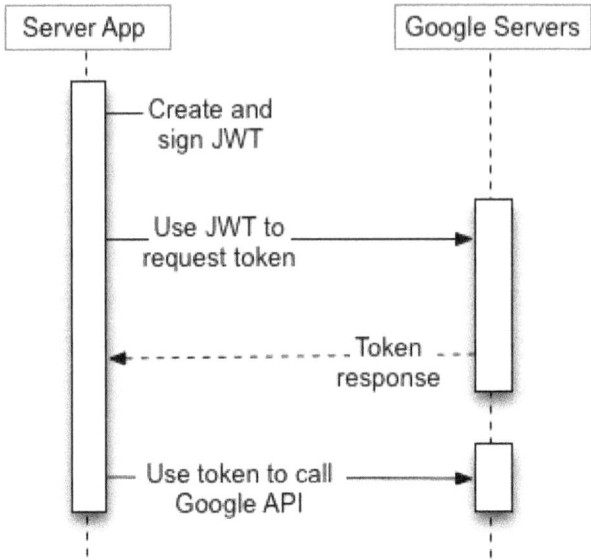

Figure 5-3. *Using OAuth 2.0*

OAuth 2.0 can be used in different ways within GCP services. Cloud Functions and Cloud Run can be used to ensure the authentication of resources between each of these services. Also, third-party integrations can use the platform to authenticate requests going to external services and APIs. API Gateway can use the authentication model to ensure that clients can provide a valid access token. This means they get the correct information in each instance.

You can use OAuth 2.0 on GCP in different. First, scope permissions can limit permissions provided by the access token. In addition, secure storage for client access tokens and secrets is essential. Additionally, short-lived tokens ensure that there is a short expiration timeline, preventing against the misuse of privileges. You can use the following Python code to set up OAuth 2.0 for service accounts:

```
from google.oauth2 import service_account
from google.auth.transport.requests import Request

# Path to your service account key file
SERVICE_ACCOUNT_KEY_FILE = 'path/to/service-account-key.json'

# Define the scopes for the access token
SCOPES = ['https://www.googleapis.com/auth/cloud-platform']
```

```
# Create credentials using the service account key file
credentials = service_account.Credentials.from_service_account_file(
    SERVICE_ACCOUNT_KEY_FILE, scopes=SCOPES
)

# Refresh the credentials to get an access token
credentials.refresh(Request())

# Print the access token
print("Access Token:", credentials.token)
```

API Keys for Service-to-Service Authentication

API keys provide a simpler platform for authentication compared to OAuth 2.0. API keys are unique identifiers that come with requests and help identify the request caller. They are less secure and can be used only when greater security interventions have been provided. The API keys start working when keys are generated on the service provider platform, after which they are shared with the client. Additionally, the API keys are included in requests, ensuring that they are in the HTTP headers or query parameters. After the keys are included, they can be validated and identified by the service providers, which evaluate their requests, look at the permission needs, and either accept or deny their access. These keys can be used in different ways (de Almeida & Camedo, 2022). They can be used in public APIs to track and control the level of access provided for every platform and system. Additionally, they can be used with rate limits, ensuring the prevention of abuse and ensuring safety for APIs. API keys can be used for internal services, enabling strict security configuration and further communication efficiency whenever needed. API Keys can be used to secure Cloud Functions by using the following Python code:

```
import functions_framework
import os

# Environment variable for the expected API key
EXPECTED_API_KEY = os.getenv('EXPECTED_API_KEY')

@functions_framework.http
def secure_function(request):
    # Get the API key from the request headers
```

```
    api_key = request.headers.get('X-API-Key')

    # Validate the API key
    if api_key != EXPECTED_API_KEY:
        return 'Unauthorized: Invalid API key', 401

    # Proceed with the function logic
    return 'Hello, World! This is a secure function.', 200
```

API keys can be implemented in GCP in different ways. The API Gateway service supports keys in the authentication of requests, working to ensure the creation, management, and provision of API keys within GCP Console to secure services as desired. Also, API keys can work with cloud endpoints to ensure security for APIs, enhancing the results whenever needed. Using custom logic ensures that the API keys can be validated on applications. API keys requires users be restricted to specific instances where the referrers, services, and IP addresses have major security issues (De, 2023). The continuous rotation of the API keys is another instrumental step to enforce security on the system. Furthermore, monitoring and auditing API key usage helps to detect anomalies and suspicious use. You can code the storage of API keys using Google Secret Manager with the following Python code:

```
from google.cloud import secretmanager

# Initialize the Secret Manager client
client = secretmanager.SecretManagerServiceClient()

# The project ID and secret ID
PROJECT_ID = 'your-project-id'
SECRET_ID = 'your-secret-id'

# Create a secret
secret_name = f'projects/{PROJECT_ID}/secrets/{SECRET_ID}'
secret = client.create_secret(
    request={
        'parent': f'projects/{PROJECT_ID}',
        'secret_id': SECRET_ID,
        'secret': {'replication': {'automatic': {}}},
    }
)
print(f"Created secret: {secret.name}")
```

```
# Add a version to the secret (store the API key)
API_KEY = 'your-api-key'
response = client.add_secret_version(
    request={
        'parent': secret.name,
        'payload': {'data': API_KEY.encode('UTF-8')},
    }
)
print(f"Added secret version: {response.name}")

# Retrieve the API key
version_name = f'{secret.name}/versions/latest'
response = client.access_secret_version(request={'name': version_name})
retrieved_api_key = response.payload.data.decode('UTF-8')
print("Retrieved API Key:", retrieved_api_key)
```

Table 5-1 compares OAuth 2.0 and API keys.

Table 5-1. *Comparison of OAuth 2.0 and API Keys*

Quality	OAuth 2.0	API Keys
Complexity	Token management	No token management
Security	Token-based authentication	Uses simple identifiers
Expiration	Uses short-lived tokens	Long-lived tokens
Permissions	Granulated permission provision	Basic permission management and provision

Using both API keys and OAuth 2.0 can be instrumental in ensuring both internal and public communication needs. API keys apply to low-risk endpoints and can be used for public APIs that do not have additional communication needs. Additionally, using OAuth 2.0 for internal service-to-service communication is critical. Therefore, the combination can enable a successful security service provision at all times.

Securing communication channels is essential, and using extra layers like mutual TLS (mTLS) helps strengthen system security. It's also important to identify authentication issues, validate access tokens, and reject invalid API keys based on permissions. These steps support secure architecture and improve deployment. With proper authentication, monitoring, and logging, security challenges in GCP can be better managed to meet growing demands.

Service-to-service authentication can be used to secure APIs and workloads on GCP. It can provide robust, flexible security mechanisms such as handling API keys and tokens. In addition, Workload Identity Federation can help manage security for both hybrid and multicloud channels. Finally, using Identity-Aware Proxy (IAP) makes it easier to have fine-grained access policies.

III. Real-World Example: Securing an API-Based SaaS Product

The modern digital landscape requires businesses to use software to help run different services. Businesses, therefore, rely on the software-as-a-service (SaaS) model to provide these services. Securing SaaS requires several considerations.

For example, say Company X provides a software service to help organizations evaluate, diagnose, and monitor their cloud service capabilities and standards. Company X provides an API to the customers to ensure that they can generate reports, monitor metrics, and implement alert systems to help them find solutions. Therefore, Company X needs to make sure that the internal services and client organizations are well secured, calling for them to have a multilayered security approach for the system to function well.

Company X implements API Gateway to serve as an entry point, and it works with all incoming API requests. The platform is critical to ensure that API Gateway can act as a central platform for handling every API. The security is monitored using OAuth 2.0 and API keys. Rate limiting ensures that no one abuses the requests. Monitoring tools are also implemented to track API usage.

OAuth 2.0 helps authorize requests after they have been authenticated. Client registration provides each client with a secret and ID. The company uses these credentials to get tokens from authorization servers and grants API access for key system functions. This ensures requests are linked to individual clients, supporting effective development and consistent service improvement. Company X also uses API keys to help provide internal services. Monthly reports on cloud data ensures proper access to these valuable platforms. Furthermore, the API keys are regularly rotated, leading to better management of the storage on the Google Secrets Management platform.

Company X has additionally worked to ensure that the backend logic is implemented through Cloud Functions and Cloud Run. These have assigned service accounts, ensuring the minimum permissions are granted to perform the tasks. GCP has enabled Company X to have a secure API-based SaaS product.

References

Borra, P. (2024). A Survey of Google Cloud Platform (GCP): Features, Services, and Applications. *International Journal of Advanced Research in Science, Communication and Technology (IJARSCT) Volume, 4.*

de Almeida, M. G., & Canedo, E. D. (2022). Authentication and authorization in microservices architecture: A systematic literature review. *Applied Sciences*, *12*(6), 3023.

De, B. (2023). API management. In *API Management: An Architect's Guide to Developing and Managing APIs for Your Organization* (pp. 27-47). Berkeley, CA: Apress.

Guptha, A., & Murali, H. (2021, May). A comparative analysis of security services in major cloud service providers. In *2021 5th International Conference on Intelligent Computing and Control Systems (ICICCS)* (pp. 129-136). IEEE.

Kendyala, S. H. (2023). High Availability Strategies for Identity Access Management Systems in Large Enterprises. *Available at SSRN 5074869*.

Kukreti, P. (2023). *Google Cloud Platform All-In-One Guide: Get Familiar with a Portfolio of Cloud-based Services in GCP (English Edition)*. BPB Publications.

Mohammed, I. A. (2019). Cloud identity and access management–a model proposal. *International Journal of Innovations in Engineering Research and Technology*, *6*(10), 1-8.

Muppa, K. R. (2024). Study on Cloud-Based Identity and Access Management in Cyber Security. *International Journal of Data Analytics Research and Development (IJDARD)*, *2*(1), 40-49.

Roy, A., Banerjee, A., & Bhardwaj, N. (2021). A study on google cloud platform (gcp) and its security. *Machine Learning Techniques and Analytics for Cloud Security*, 313-338.

Triartono, Z., & Negara, R. M. (2019, September). Implementation of role-based access control on OAuth 2.0 as authentication and authorization system. In *2019 6th international conference on electrical engineering, computer science and informatics (EECSI)* (pp. 259-263). IEEE.

Vasudevan, A. (2023). Formal Analysis and Verification of OAuth 2.0 in SSO.

Wilson, Y., & Hingnikar, A. (2022). Oauth 2 and api authorization. In *Solving Identity Management in Modern Applications: Demystifying OAuth 2, OpenID Connect, and SAML 2* (pp. 63-101). Berkeley, CA: Apress.

CHAPTER 6

Automating IAM Policies

I. Using Terraform to Manage IAM at Scale

As enterprises grow, they need increasingly better security to manage their cloud environments. To meet this need, Identity and Access Management (IAM) policies must be applied at scale. Infrastructure as code (IaC) tools such as Terraform can help organizations manage IAM more effectively, allowing them to scale their systems to meet their goals.

Terraform can define the cloud infrastructure with regard to IAM policies and offers management tools. Terraform can ensure the declarative use of IAM policies and enhance security. Terraform has these benefits:

- **Modularity:** Terraform is made up of different modules, ensuring that there are composable and reusable components of the IAM policies, roles, and permissions at all times. This modularity offers a lot of consistency when enforcing policies in multiple environments.

- **Declarative syntax:** Terraform uses a declarative configuration language, which helps to define the infrastructure.

- **State management:** Terraform has state files that can track the current status of infrastructure at any given instance. This means the IAM resources are in sync with the configurations (Shirinkin, 2017).

- **Version control:** Terraform ensures that different configurations can be stored in various locations such as GIT, enhances collaboration. This also means there are rollback capabilities and audits that can be done.

- **Cross-platform Support:** Using Terraform enables enterprises to configure multiple cloud providers such as Google Cloud, AWS, and Azure.

To use Terraform, the organization has to set up the environment correctly. An organization has to download and install Terraform. The next steps are to configure the cloud service provider and set up the credentials on GCP (Salecha, 2022). The last step is to initialize a Terraform project, which can be done on the Terraform configuration.

Setting Up Terraform for GCP IAM Management

The first step is to configure the GCP provider, including defining the GCP project and necessary credentials. The GCP configuration can be configured using the following code:

```
provider "google" {
  project = "your-gcp-project-id"
  region  = "us-central1"
  zone    = "us-central1-a"
}
```

The next step is to authenticate Terraform with GCP so it can manage the resources. The service accounts need the right permissions to download the JSON file. The service account is set within the IAM & Admin section of the GCP console, under the service accounts, where a new service account is created (Mahjourighasroddashti, 2023). A role is then assigned as roles/resourcemanager.project, which specifies the permissions that the service account will have. The next step involves downloading JSON key file and setting the environment variable GOOGLE_APPLICATION_CREDENTIALS to point to the JSON file as follows:

> *export GOOGLE_APPLICATION_CREDENTIALS="path/to/your/service-account-key.json"*

After compiling the configuration, the next step is to configure Terraform to help with the necessary plugins. This is done with the following:

> *terraform init*

The next step involves defining the policies for each user. You can use the following code to automate the policy definition:

```
resource "google_project_iam_policy" "project" {
  project     = "your-gcp-project-id"
  policy_data = data.google_iam_policy.admin.policy_data
}
```

```
data "google_iam_policy" "admin" {
  binding {
    role = "roles/viewer"
    members = [
      "group:viewers@your-domain.com",
    ]
  }
}
```

    This code defines the permissions for members. Defining roles in Terraform is critical to achieve reliable outcomes. After defining the IAM policies, you can apply these policies automatically with the command *terraform apply*. Terraform will produce a plan of the changes to be made to be evaluated. The next step is to enable the IAM policies at scale (Valtanen, 2023). This results in the same policies being used on several resources and multiple projects. The Terraform modules encapsulate configurations and make it possible to reuse them. You can create an IAM module with the following code:

```
# modules/iam/main.tf
resource "google_project_iam_policy" "project" {
  project     = var.project_id
  policy_data = data.google_iam_policy.admin.policy_data
}

data "google_iam_policy" "admin" {
  binding {
    role    = var.role
    members = var.members
  }
}

# modules/iam/variables.tf
variable "project_id" {
  description = "The GCP project ID"
  type        = string
}

variable "role" {
  description = "The IAM role to assign"
```

Chapter 6 Automating IAM Policies

```
  type        = string
}

variable "members" {
  description = "The members to assign the role to"
  type        = list(string)
}
```

For the main module, it can be coded as:

```
module "project_iam" {
  source     = "./modules/iam"
  project_id = "your-gcp-project-id"
  role       = "roles/viewer"
  members    = ["group:viewers@your-domain.com"]
}
```

The next step is to set up the management of IAM policies for particular resources. Resource-level management can be done with the following code:

```
resource "google_storage_bucket_iam_policy" "bucket" {
  bucket      = google_storage_bucket.my_bucket.name
  policy_data = data.google_iam_policy.bucket_admin.policy_data
}

data "google_iam_policy" "bucket_admin" {
  binding {
    role = "roles/storage.admin"
    members = [
      "user:admin@your-domain.com",
    ]
  }
}

resource "google_storage_bucket" "my_bucket" {
  name     = "my-unique-bucket-name"
  location = "US"
}
```

CHAPTER 6   AUTOMATING IAM POLICIES

This code manages IAM policies for a GCS bucket, allowing appropriate access control to support safety management needs. The binding block specifies which users can access the bucket, reinforcing policy enforcement.

To ensure that the IAM management can scale, you can use Terraform states (Talluri, 2023). Terraform states ensure that the multiple project sandbox environments using this particular resource can be configured to have the right IAM policies. The configuration of the GCS bucket is as follows:

```
terraform {
  backend "gcs" {
    bucket  = "your-terraform-state-bucket"
    prefix  = "terraform/state"
  }
}
```

Remote state storage ensures that the Terraform state can be securely stored and accessible by the enterprise's team. The last step is to use CI/CD to deploy the IAM policies. The automation ensures that Terraform is integrated into the CI/CD pipeline to ensure that changes to the configurations are received across the board (Cabianca, 2024). This can be coded as follows:

```
name: 'Terraform'

on:
  push:
    branches:
      - main

jobs:
  terraform:
    runs-on: ubuntu-latest
    steps:
      - name: Checkout code
        uses: actions/checkout@v2

      - name: Set up Terraform
        uses: hashicorp/setup-terraform@v1
```

```
- name: Terraform Init
  run: terraform init

- name: Terraform Apply
  run: terraform apply -auto-approve
```

This code enables an automation service.

To get all the benefits of Terraform, best practices should be followed such as configuring IAM roles using the policy of least privilege. This theory means providing a service account with the least amount of permissions necessary to handle their needed tasks. Additionally, using variables and outputs will ensure that Terraform is reusable and flexible (Verdet et al., 2025). In addition, implementing auditing and monitoring creates a chance to evaluate unauthorized changes and policy issues. The approach gradually builds the ability to review and plan changes, ensuring IAM policies are properly considered for long-term, sustainable engagement (see Figure 6-1). It also requires documenting policies to provide clear guidance for IAM configurations on the Terraform platform.

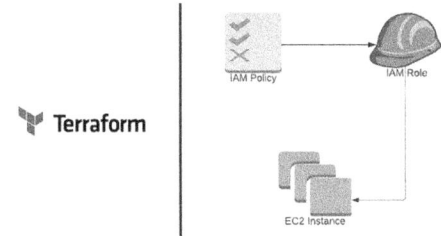

Figure 6-1. *IAM policies through Terraform*

Managing IAM policies at scale is complex in GCP. Terraform, however, offers the ability to automate the process. Terraform works well for this because of its modular design and declarative approach, making it an ideal tool to handle automation.

II. Automating Workflows with Scripts and the IAM API

Even with Terraform's potential, the process of automation in some instances requires direct interaction and scripting with the IAM API. Using scripts can improve the IAM workflows within GCP to ensure the best possible service provisions at all times.

The IAM API manages the IAM policies of GCP resources. It is a RESTful API with endpoints that can cater to various needs. These endpoints can assist in managing predefined roles and custom roles, ensuring that a user can create and update the roles to meet their needs (Nawaz, 2024). The roles can also be applied to test permissions on resources, enabling an organization to learn about their permission allocations and the limits within the organization. Finally, they can be used to get and set IAM policies on folders, projects, and organizations. They can also be used for specific resources such as the Cloud Storage buckets. Therefore, the IAM API is key to the GCP API ecosystem because it can interact with command-line tools, client libraries, and direct HTTP requests.

Before doing any scripting, make sure to meet the following requirements:

- **GCP project:** A GCP project is required with the right IAM settings to help manage policies.

- **Programming language:** You have to be familiar with a programming language to do any scripting. Languages such as Bash, Python, and Go are supported.

- **Service account:** A service account is needed with the necessary permissions. Permissions such as roles/iam.securityAdmin are required.

- **IAM API enabled:** The project should have been enabled for the IAM API, ensuring that they can mitigate different strategies for managing and handling IAM policies.

Furthermore, the IAM API has to be authenticated, and the authentication with the service account has to be approved. The service account can be created using the GCP console, and the JSON key is downloaded as shown in Figure 6-2. The next step is setting the environment variable, which includes the service account. The client library is then installed in the desired programming language such as Python.

```
{
    "type": "service_account",
    "project_id": "d112324-163219",
    "private_key_id": "asdf",
    "private_key": "-----BEGIN PRIVATE KEY-----\nMIIEvgIBADANBgk
    "client_email": "myapp-drive-only@datadocs-163219.iam.gservi
    "client_id": "115680643234234697815843778",
    "auth_uri": "https://accounts.google.com/o/oauth2/auth",
    "token_uri": "https://oauth2.googleapis.com/token",
    "auth_provider_x509_cert_url": "https://www.googleapis.com/c
    "client_x509_cert_url": "https://www.googleapis.com/robot/v1
}
```

Figure 6-2. Service account JSON key

When working with resources, automated services can handle different functions, such as to retrieve IAM policies (Ivanov, 2021). The ability to retrieve an IAM policy helps with auditing and debugging. The following code shows how to retrieve the policies:

```
from google.cloud import iam
from google.oauth2 import service_account

# Authenticate using the service account key
credentials = service_account.Credentials.from_service_account_file(
    "path/to/your/service-account-key.json"
)

# Initialize the IAM client
client = iam.IAMClient(credentials=credentials)

# Define the resource (e.g., a project)
resource = "projects/your-gcp-project-id"

# Get the IAM policy
policy = client.get_iam_policy(request={"resource": resource})

# Print the policy
print(policy)
```

This code retrieves an IAM policy for a particular GCP project, further ensuring that it can be presented to the console. Scripts can also be used to update IAM policies. The following is the code for automating the update process:

```python
from google.cloud import iam
from google.oauth2 import service_account
from google.iam.v1 import policy_pb2

# Authenticate using the service account key
credentials = service_account.Credentials.from_service_account_file(
    "path/to/your/service-account-key.json"
)

# Initialize the IAM client
client = iam.IAMClient(credentials=credentials)

# Define the resource (e.g., a project)
resource = "projects/your-gcp-project-id"

# Get the current IAM policy
policy = client.get_iam_policy(request={"resource": resource})

# Add a new binding for the 'roles/viewer' role
new_binding = policy_pb2.Binding(
    role="roles/viewer",
    members=["group:viewers@your-domain.com"]
)
policy.bindings.append(new_binding)

# Update the IAM policy
client.set_iam_policy(request={"resource": resource, "policy": policy})

print("IAM policy updated successfully.")
```

You can also create custom roles with scripts to increase efficiency and assign permissions. The following is the code to create a custom role:

```python
from google.cloud import iam
from google.oauth2 import service_account

# Authenticate using the service account key
credentials = service_account.Credentials.from_service_account_file(
    "path/to/your/service-account-key.json"
)
```

```
# Initialize the IAM client
client = iam.IAMClient(credentials=credentials)

# Define the custom role
custom_role = {
    "title": "Custom Storage Viewer",
    "description": "Allows viewing storage buckets but not
    modifying them.",
    "included_permissions": ["storage.buckets.get", "storage.
    buckets.list"],
}

# Create the custom role
response = client.create_role(
    request={"parent": "projects/your-gcp-project-id", "role": custom_role}
)

print("Custom role created:", response.name)
```

Automating IAM workflows is beneficial way to meet the needs of an enterprise in managing their policies and permissions. There are various practices that can be used to ensure that you are making valuable additions in the IAM workflows. Primarily, following the least privilege method means having the minimum permissions needed to ensure the tasks are completed. Monitoring/auditing the IAM permissions also enhances the workflow because it's an efficient way to detect errors and fix them before they affect the system. Version control offers the ability to store scripts and ensure the changes can be tracked. Version control also improves team collaboration, creating sustainable value for the AIM workflows (Sankaran et al., 2020). In addition, documentation marks the chance to explain every element of the workflow. Storing these scripts is a way to articulate valuable additions to the system. The error handling approach ensures edge cases are properly managed, helping IAM workflow automation succeed and support better organizational development.

The process of automating IAM workflows with scripts alongside the IAM API ensures flexibility and critical management advancement for permissions at scale, as indicated in Figure 6-3. GCP can automate retrieving policies, testing permissions, and creating custom roles. Using a programming language like Python on Gcloud can streamline IAM management and enhance security as it reduces chances of human error in managing their required advances (Ramuka, 2019). These advanced techniques assist in automating IAM policies.

AWS IAM Workflow

How Does IAM work?

Core IAM components
- A database containing users' identities and access privileges.
- IAM tools for creating, monitoring, modifying, and deleting access privileges.
- A system for auditing login and access history

Identify

Authenticate

Authorize

cloudanix

Figure 6-3. IAM workflows

III. Real-World Governance Strategies for Large Enterprises

Cloud resources in large enterprises require compliance with both government and organizational policies. As organizations scale their cloud infrastructure, strong governance is essential to ensure security, efficiency, and compliance. The following areas must be properly addressed to manage GCP effectively and maintain high standards:

i. **Centralized Identity and Access Management (IAM):**
   Organizations need to have a centralized IAM system. The centralization of IAM enables effective governance, as the access to organizational cloud resources is consistently handled throughout the organization, decreasing instances of unauthorized access. Organizational-level IAM policies ensure centralization. Organizations have the capacity to define IAM

policies across project, folders and organization levels. Setting these policies will ensure that organizations have an increasingly beneficial way to handle access across every project and folder. Additionally, organizations can use RBAC to designate key roles and activities that have to be conducted by every user. This defines the permissions for each predefined role. Finally, using service accounts for automation is a key step to ensure safety. The service accounts can be granted the minimum permissions and have regular audits, leading to a safe and practical way to govern the permissions on the different organizational levels.

ii. **Using organizational policies to enforce policies:** Organization policies in GCP help define constraints on resources, ensuring governance and compliance are maintained for better management. This can be done by restricting the resource locations (Qiu et al., 2020). The resource locations can be restricted based on where they are created, reducing accidental deployments. This marks the chance to work only within defined regions. Organizations can additionally work with security best practices. Organizations can also limit API access when modeling their access engagement models.

iii. **Auditing and monitoring:** Organizations can enable auditing and monitoring through Cloud Audit logs. This means they can collect information, track changes, and investigate incidents as they occur. Using the GCP security command center will improve the security posture and help locate misconfigurations, vulnerabilities, and compliance violations. In addition, setting up alerts and notifications will ensure administrators have the necessary information about any suspicious activities and policy violations that are yet to occur.

iv. **Automation for governance:** Automation helps companies scale by enforcing policies, managing resource deployment, and ensuring compliance. Automated checks and remediation can identify and address underlying issues.

v. **Multitenant environments:** Organizations, when operating multitenant environments, have to outline access controls and exercise clear isolation. Using folders can assist in isolation and ensure that IAM policies enhance access controls. This can be conducted on the department, business unit, or team levels. Resource quotas help prevent teams from overusing resources, supporting better regulation and management. Organizations should also monitor cross-tenant access to ensure users don't have excessive permissions.

vi. **Compliance and regulatory requirements:** Big organizations have to comply with a wide range of compliance rules such as PCI DSS, HIPAA, and GDPR. GCP offers various tools to help handle these regulatory needs. It can encrypt data at rest and in transit. Access controls regulate the access to data based on roles and responsibilities, ensuring that these compliance settings are properly handled. Finally, regular audits of the system will reveal any changes have to be made.

vii. **Training and awareness programs:** Organizations have to train their employees on cloud network security. They need to implement awareness programs about the risks, threats, and compliance needs; this will help employees and management understand how to secure their systems. Collaborating with internal teams and external organizations on security helps improve understanding of current and future risks. These efforts also help identify and address security issues across all levels of the organization.

Implementing effective governance ensures large enterprises can safeguard against risks in the cloud environment. The centralization of IAM, audits, task automation, and security compliance are key to advancing solutions to critical policy engagement dimensions. Thus, organizations have to continually work toward advancing safe practices even with automation at scale. Using platforms such as Terraform will help organizations deploy automated IAM policies at scale, leading to better management.

IV. Pro Tips for Automating Policy Updates Efficiently

To ensure an advanced and critical automation of the policy updates, an organization can implement different approaches. Each of the following approaches will offer benefits:

   i. Use infrastructure as code tools such as Terraform and Pulumi.

   ii. Using GCP organizational policies to enhance automation.

   iii. Use Open Policy Agent to implement policy as code.

Each of these interventions is critical for meaningful IAM policies within an organization. Companies can use these strategies to best use their automated policy update systems.

References

Cabianca, D. (2024). Configuring Access. In *Google Cloud Platform (GCP) Professional Cloud Security Engineer Certification Companion: Learn and Apply Security Design Concepts to Ace the Exam* (pp. 15-175). Berkeley, CA: Apress.

Ivanov, O. (2021). Development of CI/CD platform deployment automation module for group software development.

Mahjourighasroddashti, A. (2023). Automation of Project Creation on Google Cloud.

Nawaz, M. K. (2024). Exploring end-to-end data engineering a GCP case study.

Qiu, J., Tian, Z., Du, C., Zuo, Q., Su, S., & Fang, B. (2020). A survey on access control in the age of internet of things. *IEEE Internet of Things Journal, 7*(6), 4682-4696.

Ramuka, M. (2019). *Data analytics with Google Cloud platform.* BPB Publications.

Salecha, R. (2022). Authentication and Authorization. In *Practical GitOps: Infrastructure Management Using Terraform, AWS, and GitHub Actions* (pp. 323-395). Berkeley, CA: Apress.

Sankaran, A., Datta, P., & Bates, A. (2020, December). Workflow integration alleviates identity and access management in serverless computing. In *Proceedings of the 36th Annual Computer Security Applications Conference* (pp. 496-509).

Shirinkin, K. (2017). *Getting Started with Terraform.* Packt Publishing Ltd.

Talluri, S. (2023). Saviynt Meets GCP: A Deep Dive into Integrated IAM for Modern Cloud Security. *Journal of Information Security, 15*(1), 1-14.

Valtanen, V. (2023). IMPROVING THE MAINTAINABILITY AND DEVELOPER EXPERIENCE OF TERRAFORM CODE.

Verdet, A., Hamdaqa, M., Silva, L. D., & Khomh, F. (2025). Assessing the adoption of security policies by developers in terraform across different cloud providers. *Empirical Software Engineering, 30*(3), 74.

CHAPTER 7

Auditing and Monitoring IAM Policies

I. Using Cloud Audit Logs to Track IAM Policy Changes

Organizations can use Identity and Access Management (IAM) policies to define access in cloud environments. They can use IAM to manage and monitor their IAM policies at all times. Governing IAM policies using Cloud Audit Logs is an instrumental step when handling changes to IAM policies.

Understanding Cloud Audit Logs

The Cloud Audit Logs tool logs events occurring within the cloud environment. The logs contain information about different activities such as access attempts, configuration changes, and API calls. The audit logs offer insight into who made changes, what specific changes were made, and the timing of these changes. The majority of cloud service providers offer built-in logging capabilities, further enhancing the potential for the clients to continuously assess activities within cloud platforms (Roy et al., 2021). GCP provides Cloud Audit Logs, as indicated in Figure 7-1, to provide insight into Admins Activity logs and data access logs. These logs contain information about the events occurring, and look into metadata of resources, and help craft the best response to the cloud environment activities.

CHAPTER 7   AUDITING AND MONITORING IAM POLICIES

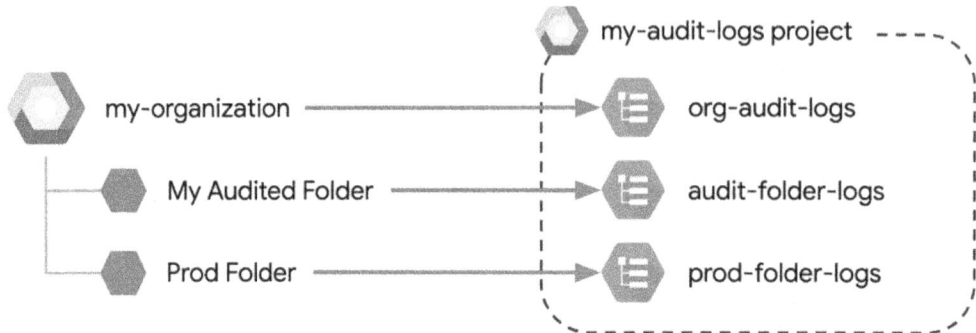

Figure 7-1. *Google Cloud Audit Logs*

Enabling Cloud Audit Logs

The process of enabling Cloud Audit Logs begins with accessing the Cloud Console on GCP and going to the Logging section, which shows the key installations of the cloud audits. The next step is to configure the log storage function and ensure that the storage is happening within the GCP cloud storage bucket. The next step is to identify which logs need to be retained by aligning with the organization's compliance requirements and managing log retention effectively. Advanced steps include enabling logging for IAM policies, tracking changes and developments, and maintaining a clear record of IAM policies that must be highlighted whenever changes occur.

Tracking IAM Policy Changes with Cloud Audit Logs

After enabling Cloud Audit logs, you need to configure how IAM policy changes happen, including logging demands. You first need to identify the relevant log entries and ensure there are key entries that are recorded to avoid collecting too general information. Collecting information helps capture key events like the creating, update, or deleting a policy. These entries provide a way to track and review all changes. Log details are then analyzed to ensure every action taken is clearly recorded. (Cambric & Ratemo, 2021). Key information includes actor, timestamp, action, request parameters, and resource used in the entire process to achieve desired actions. Each of these details is critical for achieving relevant log management needs.

More to the point, to enable successful tracking, the logs have to be correlated with IAM policies, ensuring that they have a comprehensive understanding of policy changes whenever they occur. The policies enable critical information to be displayed, leading to a much better system of handling the information provided from the logs. The analysis and tracking activities ensure that any unauthorized or suspicious modifications outside of the normal policies and parameters can be highlighted within the system. The tracking leads to a record of these detected changes and a reliable way to sort and manage the logging challenges within an organization (Kukreti, 2023). The tracking of log activities through Cloud Audit can help to generate reports and alerts. The reports on IAM policy management, demands, and progression can dictate how to make logging successful. Tracking the IAM policy changes can be conducted through the following code:

```python
from google.cloud import logging_v2

def analyze_iam_changes(project_id, time_window="1d"):
    """Detects IAM policy changes in GCP audit logs"""
    client = logging_v2.LoggingServiceV2Client()
    log_filter = (
        f'resource.type="iam_role" AND '
        f'protoPayload.methodName="SetIamPolicy" AND '
        f'timestamp>="{time_window}"'
    )

    entries = client.list_log_entries(
        resource_names=[f"projects/{project_id}"],
        filter_=log_filter,
        order_by="timestamp desc"
    )

    results = []
    for entry in entries:
        results.append({
            'timestamp': entry.timestamp,
            'user': entry.proto_payload.authentication_info.principal_email,
            'resource': entry.resource.labels['role_id'],
            'change': str(entry.proto_payload.request.policy)
        })

    return results
```

Best Practices in Using Cloud Audit Logs to Track IAM Policy Changes

Cloud Audit Logs work in different ways, each enabling the progressive modeling and management of IAM policy changes on cloud platforms. To get the appropriate performance, the following practices have to implemented:

   a. **Principle of least privilege:** Enterprises have to ensure that service accounts and users have the minimal number of resources that will help them do their jobs. Following this principle will help to avoid unauthorized IAM policy changes and access.

   b. **Centralized log management:** Enterprises have to enable a central location for auditing logs. This has to be implemented in both hybrid and multicloud environment settings. The approach makes it easier to analyze, track, and monitor IAM policy changes throughout the organization.

   c. **Regular review of audit logs:** Organizations have to review logs to enable critical perspective on IAM policy changes. This makes it simpler to address imminent issues before they have a full impact on the organization. Creating these channels helps the company meet its operational needs and adapt to the changes it constantly requires. (Estrin, 2022).

   d. **Automation:** Using a tool to do auditing makes it much easier Security information and event management (SIEM) tools automate the detection of unauthorized or suspicious IAM policy changes, helping the organization improve and shape its IAM strategy to better meet critical needs.

   e. **Using multifactor authentication:** Installing MFA on enterprise systems will ensure that only users with privileges will have the capacity to modify IAM policies. This bit ensures an additional layer of security and reduces risks associated with unauthorized entry and changes to the system.

f. **Educating to teams and employees:** Enterprises have to train their teams on how to monitor IAM policy changes. This approach will ensure critical engagement and development of IAM modeling in the organization (Murthy et al., 2024).

The process of tracking IAM policy changes is key to ensuring cloud security. Using Cloud Audit Logs provides an easier way to do the configuration. It will manage audit logs, mark any modifications, and address unauthorized changes. Doing regular audits, using MFA, and having a centralized management system are all beneficial techniques.

II. Setting Up IAM Alerts with Cloud Monitoring

Cloud security has several features to offer an organization. IAM enables access to devices and users, but you have to have the right policies and settings. The alerts can be coded as follows:

```python
from google.cloud import monitoring_v3

def create_iam_alert_policy(project_id, notification_channel):
    """Creates alert for privileged IAM changes"""
    client = monitoring_v3.AlertPolicyServiceClient()

    policy = {
        "display_name": "High-risk IAM Change Alert",
        "conditions": [{
            "display_name": "IAM Policy Modification",
            "condition_threshold": {
                "filter": 'metric.type="logging.googleapis.com/user/iam-policy-changes"',
                "comparison": "COMPARISON_GT",
                "threshold_value": 0,
                "duration": {"seconds": 60},
                "aggregations": [{
                    "alignment_period": {"seconds": 60},
                    "per_series_aligner": "ALIGN_RATE"
                }]
            }
```

```
    }],
    "notification_channels": [notification_channel]
}

return client.create_alert_policy(
    name=f"projects/{project_id}",
    alert_policy=policy
)
```

On GCP, Cloud Monitoring resources can track IAM-related events, triggering alerts and ensuring that specific conditions are noted, as indicated in Figure 7-2. Therefore, you have to use the right approach when setting up cloud monitoring (Verginadis, 2023). Thus, using Cloud Monitoring works great with IAM features, each helping to meet the needs of the organization.

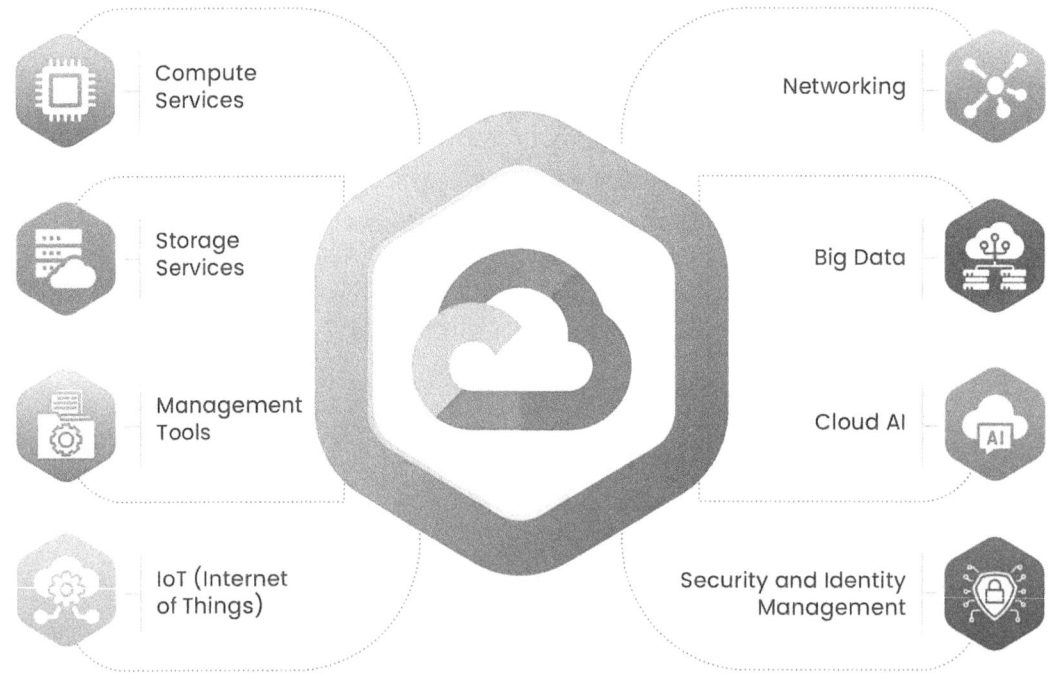

Figure 7-2. Cloud monitoring on GCP

Cloud Monitoring for IAM

GCP offsets up the cloud monitoring to collect logs, metrics, and events relating to cloud resources. This provides the ability to perform real-time monitoring and alerting. This supports different logs and information such as Data Access Logs and Admin Activity logs (Borra, 2024). Using cloud monitoring and logging allows an enterprise to set up custom IAM policy alerts, helping the organization stay responsive and secure at the same time.

Requirements for Setting Up IAM Alerts

When setting up IAM alerts, there are different tools that can be used. For example, the Cloud Monitoring API can monitor activities and handle log entries. IAM permissions have to be enabled to meet the organization's needs (Mulder, 2024). Notification channels have to be activated on the cloud platform, working to send alerts in a timely manner. Additionally, cloud logging exports have to be enabled to achieve the IAM goals.

Process of Setting Up IAM Alerts in Cloud Monitoring

When setting up IAM alerts, you want to make sure a resource can collect, evaluate, and monitor the information well. Here are the key steps needed to set up IAM alerts:

1. **Identifying critical IAM events to monitor:** This step locates high*risk functions on the system. Key actions to be considered include creating and deleting service account keys, role assignments, unexpected privilege escalations, and policy modifications.

2. **Creating log-based metrics to be used in IAM events:** Using log-based metrics will help to extract information on Cloud Audit Logs. The metrics can be used to assess policy changes and categorize activities (Kanth, 2024).

3. **Alerting policy in cloud monitoring:** When creating a policy, make sure to create conditions, collect alerts, locate triggers, and ensure messaging systems to enable the complete management of the policy.

4. **Testing the alerting strategy:** The alerting system has to be tested to make sure it is secure. Manual triggers are a great way to test the roles and advanced needs of the system. The system alerting system is verified on the notification channel to make sure every variable on the system is correct.

Advanced Alerting Strategies

Besides the normal alerting models of using IAM policies and setting metrics, different techniques can be used to implement alerts and provide information about what is happening on the system. The realistic advances in managing alerts can be conducted by merging cloud network provider designs and alerting models working best for the organization. The following are key alerting strategies:

1. **Multiproject IAM monitoring:** In organizations with multiple projects, there are different strategies that can be put into action. Aggregate logs can be critical to central logging projects. The use of log sinks is also key to forward IAM logs to the BigQuery resource to be used for analysis purposes, as indicated in Figure 7-3. Cross-project policies can be used for alerts too.

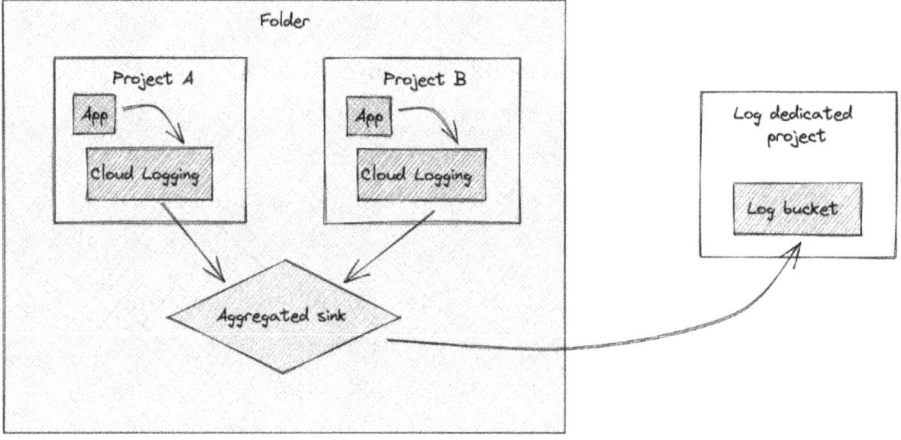

Figure 7-3. Multiproject IAM monitoring

2. **Automated remediation with Cloud Functions:** Automated remediation using cloud functions should handle critical IAM changes by triggering actions whenever an alert is generated (Valleru, 2024).

3. **Real-time anomaly detection:** Machine learning can be used to create an alerting system to identify anomalous activities. This strategic approach ensures effective handling of alerts and notifications, leading to better results.

Best Practices to Be Used in IAM Alerting Activities

IAM alerting needs to be conducted within an organization to help manage cloud security. Thus, best practices that can assist with IAM alerting include the following:

- **Employing granular filters:** This action ensures there are refined activities on log queries so that the admin users are exempt from whatever activities have to be conducted on the cloud platform.

- **Prioritization of high-risk actions:** This ensures there is an increased focus on privileged role assignments. This will also ensure that there are low-risk challenges that are ignored by alerts.

- **Regular review of alerting policies:** Obsolete alerts within the system should be removed. Additionally, organizational changes have to be considered (Bui, 2025).

- **Integration with SIEM tools:** SIEM tools can provide advanced analysis. Logs can be exported to Chronicle, Splunk, and Datadog to get even more analysis functionality.

Setting up IAM alerts through cloud monitoring on GCP is important. Organizations should implement anomaly detection, log-based metrics, and automated responses to better detect threats and respond to them in real time, reducing the impacts they have on the organizations. Configuring monitoring tools correctly can improve the organization's security.

III. Leveraging Cloud Asset Inventory for Compliance

When working in a cloud environment, organizations need to comply with regulations such as HIPAA, SOC 2, and GDPR. Using GCP means you have access to Cloud Asset Inventory (CAI), which can analyze, track, and ensure compliance with cloud resources. CAI acts as a centralized metadata repository, which includes information about all GCP assets. CAI can automatically check that organizations are complying with industrial standards. CAI also helps the organizations with audit reports that enforce regulatory compliance because it can detect misconfigurations in real time (Sialakshmi, 2021). SCC and CAI can also be used to implement continuous monitoring. Compliance can be leveraged using the following code:

```python
from google.cloud import asset_v1

def check_iam_compliance(organization_id):
    """Checks IAM policies against compliance standards"""
    client = asset_v1.AssetServiceClient()
    response = client.analyze_iam_policy(
        analysis_query={
            "parent": f"organizations/{organization_id}",
            "resource_selector": {
                "full_resource_name": f"//cloudresourcemanager.googleapis.com/organizations/{organization_id}"
            },
            "access_selector": {
                "permissions": ["*"]
            }
        }
    )

    violations = []
    for result in response.main_analysis.analysis_results:
        if result.analysis_state.code != 0:  # Non-compliant state
            violations.append({
```

CHAPTER 7   AUDITING AND MONITORING IAM POLICIES

```
            'resource': result.attached_resource_full_name,
            'role': result.iam_binding.role,
            'members': list(result.iam_binding.members)
        })
    return violations
```

CAI has various features that help manage compliance within the cloud security network. It can look at historical changes, IAM policies, and resource metadata. Table 7-1 summarizes that compliance features that can be used.

Table 7-1. Compliance Elements

Feature	Instance of Use in Compliance
Asset snapshots	Applicable in the tracking of historical changes for audit purposes
Export to BigQuery	Enhances the analysis of compliance violations in any scale
Security command center integration	Aids with detection and remediation of misconfigurations
Policy intelligence	Works with a great comparison of IAM policies versus industrial best practices

A cloud asset inventory is needed to ensure that the appropriate permissions are being used. Permissions such as roles.cloudasset.owner and roles/cloudasset.viewer help register and achieve peak performance on the audit platform. The next step is to enable an API through a command like this:

   gcloud services enable cloudasset.googleapis.com

The next step involves exporting asset data to ensure data analysis. CIA can transfer data to either four different platforms such as BigQuery, which works with SQL-based queries, Pub/Sub for handling real-time alerts, and Cloud Storage for handling long-term retention capacities. Exporting can be done with the following code:

```
gcloud asset export --organization=ORG_ID \
  --content-type=resource \
  --bigquery-table=projects/PROJECT_ID/datasets/DATASET_ID/tables/
TABLE_NAME
```

Defining policies and rules is a critical step to ensuring that the data complies with set standards and regulations. You can manage data using organizational policies and Security Health Analytics. The following policy instance restricts public IPs, further strengthening the entire compliance appeal:

> constraint: constraints/compute.vmExternalIpAccess
>
> enforce: true

Using CAI improves the automation features that help manage and enforce IAM policies. Various automation tools support this process, helping to address and manage the growing demands of data management (Shashi, 2024). For example, automating asset exports and analyzing data with BigQuery can build a strong information modeling strategy, as shown in Figure 7-4. Additionally, real-time monitoring through Pub/Sub platforms allows for faster detection and resolution of system issues as they arise.

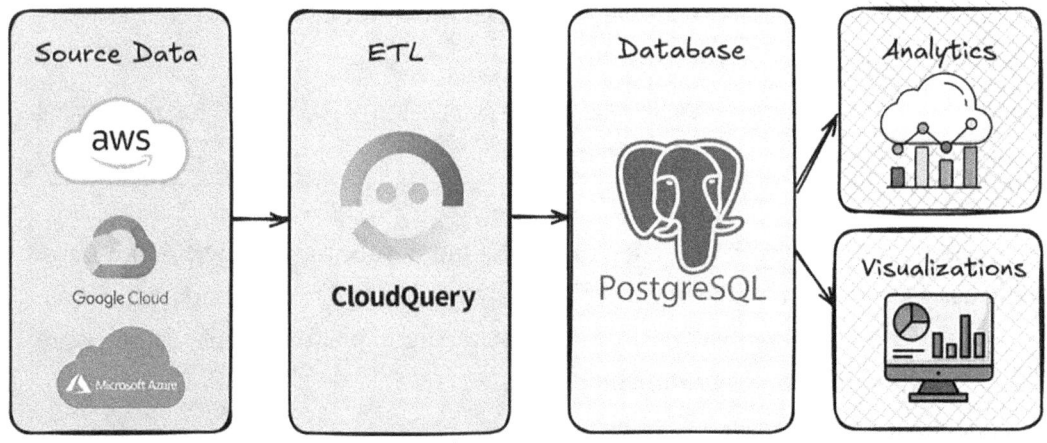

Figure 7-4. CAI function[[kim

CAI can be integrated to work seamlessly with the SCC, ensuring that the GCP has a centralized point of managing information and asserting compliance to a desired level. Additionally, the integration ensures a system of continuous compliance monitoring and handling automated misconfiguration detection, where they open firewall rules to be help in detecting any incoming and outgoing challenges. Nonetheless, the integration also helps with a prioritized risk scoring need for remediation activities.

Best Practices to Assert Compliance with CAI

Ensuring compliance with CAI helps create a useful way to apply practical incentives in different situations and makes the system easier to use and more effective. The following are best practices to enhance compliance with CAI:

1. **Regular export and analysis of asset data:** The organization can use a weekly snapshot to enhance audits and can set up automated BigQuery reports to achieve help compliance.

2. **Policy intelligence for IAM compliance:** This can help choose which best practice should be followed and can help detect any risky permissions that are in use within the system.

3. **Automate remediation:** Remediation can be automated by using Cloud Functions alongside CAI exports. This ensures any underlying issues on the platform are fixed automatically (Chaturvedi et al., 2024).

4. **Enforce guardrails with organization policies:** Noncompliant configurations have to be restricted, and each instance needs to meet certain needs.

5. **Use Cloud Audit Logs and CAI to monitor changes:** You can correlate asset changes to critical user activities to make forensic investigations easier.

These practices will help enforce critical policies and regulations within an enterprise, improving performance. Specifically, CAI helps enforce both GDPR and HIPAA, enabling policies, roles, and permissions within a cloud network to contend with the regulatory needs of these institutions.

CAI is a game changer when it comes to ensuring compliance within GCP. It can keep track of all auditing activities that have been conducted (Sugureddy, 2024). Additionally, it can help to formulate a system of real-time visibility of cloud resources, evaluating the performance of each. CAI can also automate compliance checks using SCC and BigQuery platforms. These practices enhance the provision of an automated remediation, scheduled reports, and seamless integration of the CAI and SCC, further stamping the potential to push for higher results in achieving compliance needs.

IV. Case Study: How a Healthcare Organization Uses IAM Monitoring to Meet HIPAA Compliance

A healthcare provider based in Florida operates on the GCP, which helps comply with the Health Insurance Portability and Accountability Act (HIPAA), ensuring the provision of strict control over protected health information (PHI). The healthcare provider faced challenges when auditing the IAM role changes to prevent privilege escalation, generate regular compliance reports from the audits, and track access patterns to patient data, including unauthorized access.

To enhance greater security, the company sought to implement IAM monitoring and auditing through Cloud Asset Inventory, Cloud Audit Logs, and Security Command Center, available on GCP.

To ensure robust IAM monitoring, the organization adopted a least privilege access model and took a zero-trust security approach. They implemented role-based access control (RBAC) and automated user access reviews to maintain strict access governance. Additionally, the organization leveraged Cloud Audit Logs to support real-time monitoring, focusing on critical events such as administrative activities, suspicious behavior alerts, and data access logs. To further strengthen their compliance posture, they utilized the Security Command Center (SCC) to perform automated compliance checks, scan IAM policies, generate compliance reports, and detect misconfigured resources.

These changes helped the organization achieve HIPAA compliance by reducing the risk of unauthorized access. This also supported ongoing eligibility for various incentives. The healthcare institution fostered a proactive approach to threat detection and effectively managed user privileges. Streamlined audits further improved their ability to meet compliance requirements and stay aligned with regulations.

In summary, the organization used regular IAM audits to help foster HIPAA compliance. The automation process reduced human error in access management and helped set up real-time alerts, which enhanced rapid response to security threats for the organization. Thus, the healthcare organization met the compliance regulations of HIPAA through a seamless CIA and SCC integration.

References

Borra, P. (2024). A Survey of Google Cloud Platform (GCP): Features, Services, and Applications. *International Journal of Advanced Research in Science, Communication and Technology (IJARSCT) Volume, 4.*

Bui, T. (2025). Real-time Data Analytic on Google Cloud: A Complete Data Pipeline from Self-Host Databases to GCP Services.

Cambric, S., & Ratemo, M. (2023). *Cloud Auditing Best Practices: Perform Security and IT Audits across AWS, Azure, and GCP by building effective cloud auditing plans.* Packt Publishing.

Chaturvedi, G. K., Shirole, M., & Shirole, U. (2024). An Analysis of Security and Privacy in Cloud and IoT. In *Cloud of Things* (pp. 67-89). Chapman and Hall/CRC.

Estrin, E. (2022). *Cloud Security Handbook: Find out how to effectively secure cloud environments using AWS, Azure, and GCP.* Packt Publishing Ltd.

Kanth, T. C. (2024a). AI-POWERED THREAT INTELLIGENCE FOR PROACTIVE SECURITY MONITORING IN CLOUD INFRASTRUCTURES.

Kanth, T. C. (2024b). AI-POWERED THREAT INTELLIGENCE FOR PROACTIVE SECURITY MONITORING IN CLOUD INFRASTRUCTURES.

Kukreti, P. (2023). *Google Cloud Platform All-In-One Guide: Get Familiar with a Portfolio of Cloud-based Services in GCP (English Edition).* BPB Publications.

Mulder, J. (2024). *Multi-Cloud Administration Guide: Manage and Optimize Cloud Resources Across Azure, AWS, GCP, and Alibaba Cloud.* Walter de Gruyter GmbH & Co KG.

Murthy, J. S., Siddesh, G. M., & Srinivasa, K. G. (Eds.). (2024). *Cloud Security: Concepts, Applications and Practices.* CRC Press.

Roy, A., Banerjee, A., & Bhardwaj, N. (2021). A study on google cloud platform (gcp) and its security. *Machine Learning Techniques and Analytics for Cloud Security*, 313-338.

Sailakshmi, V. (2021). Analysis of Cloud Security Controls in AWS, Azure, and Google Cloud.

Shashi, A. (2023). Advanced GCP Services. In *Designing Applications for Google Cloud Platform: Create and Deploy Applications Using Java* (pp. 155-180). Berkeley, CA: Apress.

Sugureddy, A. R. (2024). DATA GOVERNANCE EXCELLENCE IN THE CLOUD LEVERAGING GCP FOR ENHANCED LINEAGE AND SECURITY. *INTERNATIONAL JOURNAL OF DATA SCIENCE AND ANALYTICS (IJDSA), 2*(2), 8-19.

Valleru, V. (2024). Enhancing Cloud Data Loss Prevention through Continuous Monitoring and Evaluation.

Verginadis, Y. (2023, March). A review of monitoring probes for cloud computing continuum. In *International Conference on Advanced Information Networking and Applications* (pp. 631-643). Cham: Springer International Publishing.

CHAPTER 8

Managing Multicloud and Hybrid IAM

I. Integrating GCP IAM with AWS and Azure for Hybrid Environments

With the increasing adoption of both hybrid and multicloud environments, IAM is becoming more popular. GCP provides an array of IAM activities to secure and manage enterprise demands; however, companies have to work with either Azure or AWS, which can help handle consistency, security, and operational efficiency in multicloud and hybrid cloud environments.

When working in multicloud and hybrid cloud environments, challenges include the integration of IAM between AWS, GCP, and Azure. Primarily, all of these service providers have different IAM models with variations in administration tasks. While GCP offers roles and permissions, Azure works with Active Directory–based RBAC, and AWS works with policies to manage their cloud platforms (Vasanthi, 2024). In addition, these AWS providers have different security risks that have to be addressed. Further, the operational overhead when managing the multiple IAM resources from different providers can stress enterprises and cost money. Addressing these challenges to integration requires a unified strategy for identities in GCP IAM with either Azure of AWS.

Integration Strategies

Different strategies can be applied to assist in handling integration between different IAM platforms. These strategies include the following:

A. **Identity federation using single sign-on (SSO):** Federation is required to enable single authentication and enable access across all cloud provider platforms. In this case, GCP can be used as the identity provider (IdP), enhancing the capacity to configure settings on the AWS and Azure platforms. First you have to set up Cloud Identity or Google Workspace as the valid identity source, as indicated in Figure 8-1. You can use the Security Assertion Markup Language (SAML) 2.0 to ensure both AWS and Azure recognize GCP as the IdP. For Azure, GCP is the federated identity provider, and on AWS, GCP is the identity source, ensuring that roles can be designated and identities can be configured to correspond to every security policy. The SAML model thus enables GCP to act as the identity provider, enabling single sign-on as demanded. This can be achieved by employing third-party identity providers such as Azure AD, Ping Identity, and Okta. These platforms identify the primary IdP, ensuring that access can then be federated to the GCP. The GCP platform is configured to trust the external IdP using OpenID Connect or SAML, enabling the assigning of roles based on identities from the IdP. This is a chance to continue with the single sign-on regulation.

B. **Cross-cloud service accounts and Workload Identity Federation:** Workload Identity Federation can work with machine-to-machine authentication, ensuring that identities are secure. This can be done with GCP Workload Identity Federation with AWS, which makes it easier to implement security. AWS workloads such as Lambda and EC2 have to be allowed to impersonate service accounts on the GCP platform. This case allows federated identity permissions in GCP IAM, improving how service accounts are used. (Spinella, 2025). In addition, short-lived credentials enable secure access for AWS workloads on different GCP resources. When working with Azure, GCP identity

CHAPTER 8   MANAGING MULTICLOUD AND HYBRID IAM

federation can enable access to GCP for the Azure workloads such as AKS and VM. Assigning permissions will ensure that federated identities can gain access and work on the GCP platform, leading to benefits for access management. Thus, federated identities are a great way to implement united access for multiclouds and hybrid clouds on GCP. Workload Identity Federation can be checked using the following code:

```python
import google.auth
from google.cloud import iam_credentials_v1

def verify_workload_federation(aws_role_arn):
    credentials, _ = google.auth.default()
    client = iam_credentials_v1.IAMCredentialsClient
    (credentials=credentials)

    # Generate federated token for AWS
    response = client.generate_id_token(
        name=f"projects/-/serviceAccounts/your-sa@{project}.iam
        .gserviceaccount.com",
        audience=aws_role_arn,
        include_email=True
    )

    # Token can be used with AWS STS AssumeRoleWithWebIdentity
    return response.token

# Usage
token = verify_workload_federation("arn:aws:iam::123456789012
:role/gcp-
federated")
```

C. **Centralized IAM governance with policy as code:** Using the policy-as-code approach provides consistent management of IAM policies on Azure, GCP, and AWS. The main approach is to enable GCP organization policies, Azure Policies, and AWS SCPS to create a guardrail that defines their scope and level of compliance. Enabling tools such as Open Policy Agent (OPA) and HashiCorp

117

Sentinel conducts cross-cloud policy enforcement to meet any required regulations. Notably, GCP Access Context Manager can be used to add unified security controls to conditional access policies.

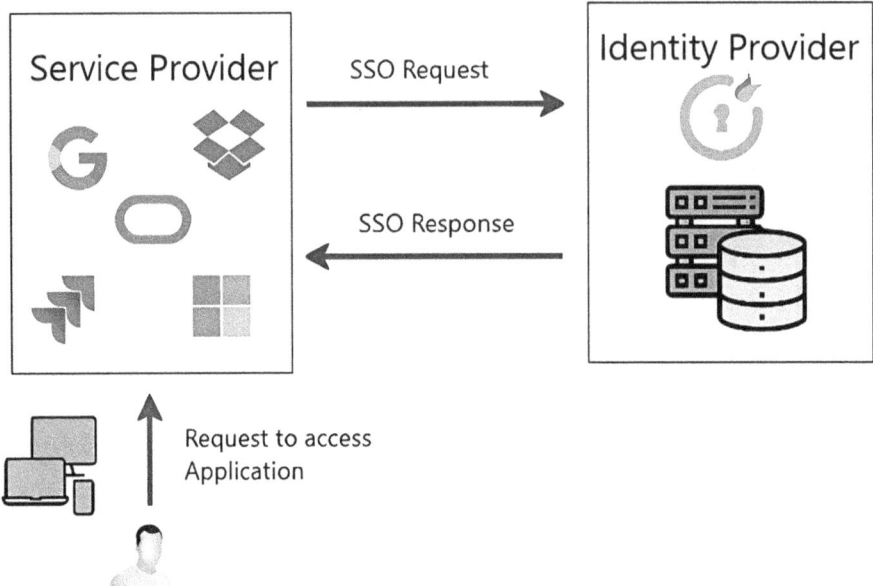

Figure 8-1. Single sign-on process

Best Practices in Securing Multicloud IAM Integration

The following are best practices when using IAM functions:

a. **Principle of least privilege:** Restricting predefined and custom roles in GCP prevents over-permissioning and supports consistent, high performance across the organization.

b. **Monitoring and auditing:** Enterprises have to constantly monitor their access patterns on different cloud platforms. This can be done using native cloud provider tools such as GCP Cloud Audit Logs, Azure Monitor and AWS Cloud Trail, providing centralization of logs for the purpose of visibility (Julakanti et al., 2022). Additionally, SIEM Tools such as Splunk and Chronicle can be used to correlate events as they occur across different cloud platforms.

CHAPTER 8 MANAGING MULTICLOUD AND HYBRID IAM

c. **User lifecycle management:** Automating user lifecycle management results in having a synchronized repository of user identities from the central directory. This also automates de-provisioning when a user leaves the organization.

d. **Use of conditional access policies:** Organizations have to continually improve the security and management their development requirements. Enforcing context-aware access will enhance security such as blocking logins from untrusted locations. The combination of tools such as Azure conditional access, AWS conditional access, and GCP BeyondCorp will secure the platform even further.

The integration of GCP IAM with cloud provider platforms such as Azure and AWS in multicloud and hybrid cloud environments is a best practice (Sidharth, 2021). The combined use of federation, centralized governance, and workload identity simplifies access management, critical compliance appeal, and enhanced security across different platforms, as indicated in Figure 8-2. Using the appropriate integration tools on multicloud and hybrid platforms will improve security and therefore engagement.

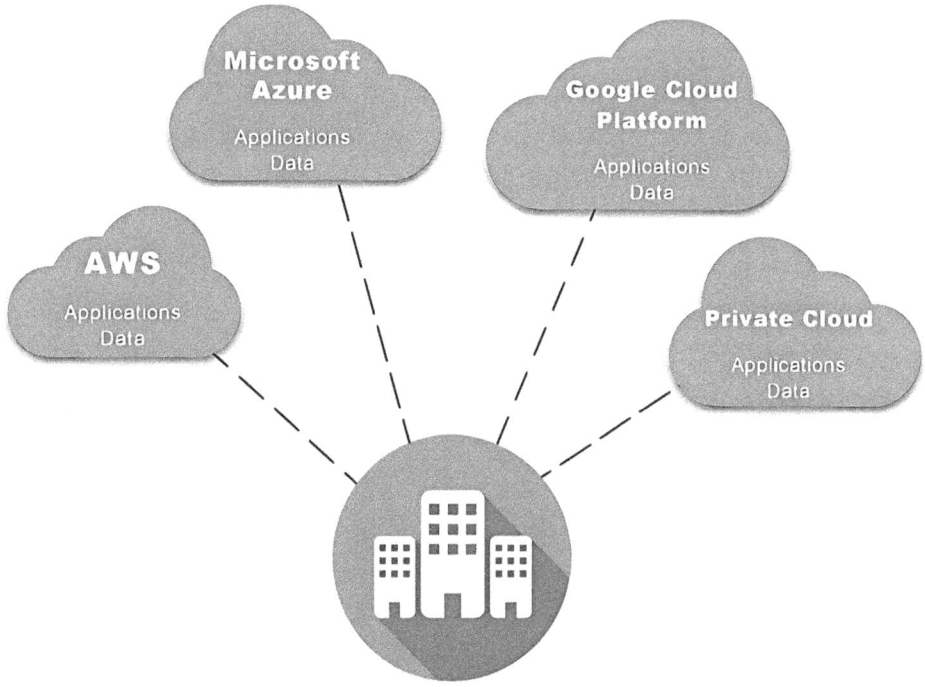

Figure 8-2. *Multicloud IAM integration*

II. Best Practices for Cross-Cloud IAM Governance

When managing a cross-cloud IAM system, you need to be aware of the best practices to achieve optimal results.

Implementation of Centralized Identity Provider (IdP)

When working with multiple cloud providers, having separated identities on Azure, GCP, and AWS often leads to credential sprawls, which can create security gaps. Using a centralized IdP provides a single source of identities, which is easier to manage (Yeboah-Ofori et al., 2024). This can be done in different ways, starting by synchronizing user attributes to ensure that the roles, groups, and attributes are consistent across all platforms. Using SAML and OICD federation creates a single sign-on capability, leading to much better engagement with different platforms. Finally, automating the provisioning and de-provisioning of roles ensures that users do not stay longer on the platforms than they are required to.

Principle of Least Privilege

Ensuring there are no over-permissioned accounts is a critical step because they can lead to security breaches. of the principle of least privilege means creating a system where users and services can only work with the required permissions that they need to achieve their tasks. Predefined roles on GCP, Azure, and AWS can be used to implement the principle of least privilege. Advancing just-in-time access is also critical to ensuring that the least privilege is addressed. Regular audits and reviews can even help come up with the best response.

Standardization of IAM Policies on Different Cloud Platforms

Each cloud provider comes with key policies that serve them in providing and advancing the right framework of control. Each of these platforms work well in advancing, preparing, and ensuring that GCP, AWS, and Azure are collectively secure. Therefore, enterprises have to standardize regulations on these platforms, enabling them to have a critical capacity to work with policies. Policy as code can ensure standard regulation, enforce governance, and make management earlier. Basically, leveraging cloud-specific policy tools can help create critical responses.

Monitoring and Auditing Cloud Access Practices

You need visibility into the access patterns and practices in order to detect policy violations and breaches whenever they occur. You can do this by ensuring a central aggregation of the logs using SIEM so they can be easily analyzed (Ike et al., 2025). Having an alert system for anomalies improves management and helps deliver better results for the platform. It also enhances the ability to review access effectively, making it easier to manage and take the right steps when working with IAM across different cloud platforms.

Secure Service Accounts and Workload Identities

Having static keys on service accounts makes them highly vulnerable to attacks, so you need extra secure authentication approaches. To enhance security, enterprises have to avoid the use of long-lived credentials on their platform and work with key rotation strategies. Enterprises have to restrict cross-cloud service account usage to limit exposure.

Emergency Access Planning Activities

Enterprises have to take measures to secure access even when the IdP approach fails. This can be done by having well-defined break-the-glass procedures that specify the emergency accounts and ensure there are credentials stored in the hardware security module to assist with handling failovers. Enterprises also need critical tests to evaluate their failover access models, leading to better results. Simulating outages is a good way to find and address emerging issues, ensuring that the organization stays secure at all times.

Automating IAM Compliance Checks

Enterprises have to move from manual compliance checks to automated checks, which are more accurate and secure. They can conduct checks by using cloud-specific compliance tools such as AWS Security Hub, GCP Security Command Center, and Microsoft Defender (Khambam & Kaluvakuri, 2023). Companies also have to automate their policy enforcement dimensions to detect and remediate policy violations quickly.

The process of enabling cross-cloud IAM governance requires the use of least privilege access, continuous monitoring, centralized identity management, policy standardization, and automation models, as indicated in Figure 8-3. These practices have to be enacted properly within an organization.

Figure 8-3. *IAM compliance checks*

III. Challenges and Solutions in Managing Multicloud Access

There are many challenges when managing multicloud access strategies in enterprises. GCP, AWS, and Azure have complexities that have to be handled, affecting flexibility, operational efficiency, and compliance needs. The challenges can affect the success of the system.

Identity Fragmentation and Siloed Access

In multicloud environments, each cloud provider has its own IAM framework designed to help manage and meet the specific requirements of their platform (Hong et al., 2019). Azure works with Azure Active Directory to ensure that it implements the role-based access control models, while AWS sets IAM policies for individual groups, users, and roles in the organization. These different instances within the multicloud environment lead to several challenges on the system, such as duplicate identities. They also use inconsistent permissions based on each user's cloud setup and rely on manual lifecycle management. These challenges affect functionality and often lead to poor outcomes on the cloud platform.

To rectify this, cloud providers have to be unified and use a common identity federation. Silos can be eliminated by having a single identity federation to enhance the capacity to handle GCP, Azure, and AWS in authenticating from a single source. This solution can be implemented using different mechanisms, such as using Workload Identity Federation and SAML/OICD-Based SSO. These avenues offer an easier way to log in from a single identity management database, resulting in better security and easier configuration.

Inconsistent Permission Models

Each cloud provider has a certain way to handle their permissions. That means to use a cloud provider platform, you have to understand how they handle permissions. GCP works with predefined roles and custom roles. The predefined roles are *roles/editor*. AWS works with JSON policies, which are attached to the roles enabled by the IAM policies. Azure works with AD roles mixed with RBAC assignments (Imram et al., 2020). This presents a series of inconsistencies on a cloud platform that might affect functionality. There may be security gaps between the cloud providers, and there might be compliance failures since the platform for checking and addressing the security needs differs from each instance of the security management. Over-permissioning is also a key challenge, which means admins have to assign broad access to users compared to whatever they actually need to work with.

The challenges can be handled by standardizing permissions. Policy as code has to enhance aspects of standardization, leading to a much better model of engaging in the needed approach. Using Open Policy Agent creates a mechanism of working with reusable policies and enforces them on GCP, Azure, and AWS (Zhao et al., 2022). HashiCorp Sentinel can also be used with the integration of Terraform to enable compliance with IAM changes. Finally, organizations can work with the policy of least privilege across different cloud provider platforms. This indicates they can use IAM Recommender on GCP to detect any instances of unused permissions, Privileged Identity Management (PIM) to enable just-in-time access, and IAM Access Analyzer to help refine policies on AWS. The scope of roles on GCP can be identified using the following Role Auditor code:

```python
from google.cloud import asset_v1

def audit_gcp_roles(project_id):
    client = asset_v1.AssetServiceClient()
    scope = f"projects/{project_id}"

    # Analyze IAM policies for custom roles with excessive permissions
    response = client.analyze_iam_policy(
        request={
            "analysis_query": {
                "scope": scope,
                "identity_selector": {
                    "identity": f"user:admin@{project_id}.iam.
                    gserviceaccount.com"
                },
                "access_selector": {
                    "permissions": ["*"]  # Wildcard for all permissions
                }
            }
        }
    )
```

```
    for result in response:
        print(f"Over-permissioned role found: {result.identity}")
        print(f"Permissions: {result.accesses}")

# Usage
audit_gcp_roles("your-gcp-project")
```

Lack of Centralized Auditing and Compliance

Using multicloud platforms has the challenge of ensuring a unified view of access logs, which affects different capabilities and duties. Security teams fail to correlate threats as they cannot access all information in a single view. In addition, the reporting for compliance is a lot of effort, since every action has to be conducted manually. The lack of a unified view also affects the ability to do detect threats in real time, where the anomalies can be different from one cloud to the next.

The challenges can be handled with unified Logging and SIEM integration. This is an approach that will aggregate cloud audit logs, ensuring an increasingly beneficial platform to handle the needs of the platform. This can be done on GCP by exporting Cloud Audit logs to Chronicle or BigQuery. For AWS, CloudTrail logs can be streamed to the S3 buckets. In Azure, it can be done by forwarding all activity logs to Log Analytics. These advances will help to aggregate the Cloud Audit logs, ensuring that they can be analyzed easily (Ramalingam & Mohan, 2021). Additionally, SIEM can be applied for the purpose of cross-cloud monitoring, ensuring that the use of platforms such as Splunk, Google Chronicle, and Sentinel are used to take in logs from all of the cloud platforms. Suspicious activities trigger alerts so that a solution can be found. The use of alerts is key to ensuring that every piece of information is processed and managed accordingly.

Managing Service Accounts and Machine Identities

Service accounts serve as a high-risk target within cloud platforms. Static keys within the service accounts are difficult to rotate, making them risky for leaking data. Nonetheless, cross-cloud access is mainly poorly managed, leading to an increased risk on the platforms and resulting in negative outcomes. These can be handled using a Workload Identity Federation approach alongside short-lived credentials. The first step is to replace static keys with federation and automate the key rotation approach. These approaches help manage the information to a desired level.

Compliance on Various Jurisdictions

For organizations in different regions, there are several challenges that have to be addressed. Different regulations have to be complied with. In the first instance, policy enforcement and data residency controls ensure that there are location-based restrictions and automated compliance checks (Patel, 2024). These advances help handle noncompliance and work toward enabling broader steps to manage any challenges to data modeling in the documented regions of the organization (Somanathan, 2023).

IAM has to be managed across multiple cloud networks by implementing some key strategies such as having a centralized monitoring approach, ensuring consistent implementation of the policy of least privilege, and working with an automated compliance mechanism, as Figure 8-4 indicates. A secure multicloud IAM strategy is therefore key to ensuring efficiency across all platforms, leading to better management.

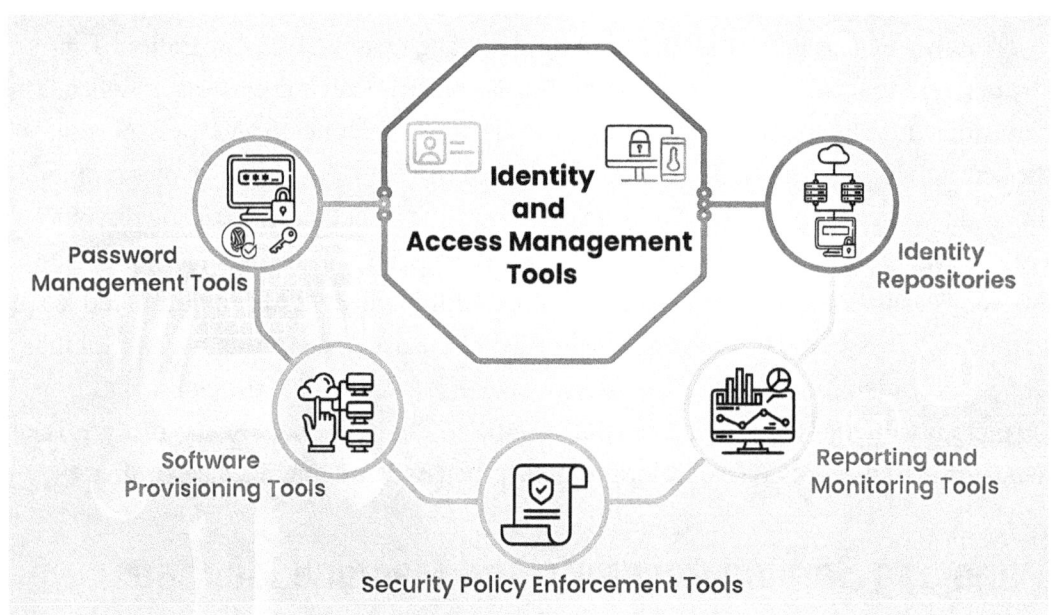

Figure 8-4. IAM compliance across jurisdictions checklist

IV. Emerging Trend: The Role of Zero-Trust Architecture (ZTA) in Hybrid Cloud Environments

The use of hybrid and multicloud environments calls for new interventions in security. The main approach to security within these cloud environments is the use of Zero Trust Architecture, which creates secure access across AWS, GCP, and Azure, even when working with on-premises systems. The use of zero trust is key to ensuring continued management of security, leading to a better system. ZTA offers benefits when there is no instance of trust, as it does not establish any implicit trust, making sure that there is no level of trust for users across the provided platforms (Ahmadi, 2024). More to the point, they allow for dynamic access control, where they work to ensure user identity, location, device health, and behavior are monitored before any form of access is provided to a user on the platform.

ZTA also works by implementing microsegmentation for workloads. This works by ensuring that there is fine-grained network segmentation, which makes it simpler to handle breaches by guarding against any lateral movements. Zero trust can be implemented in cross-cloud instances by using the following code:

```
import opa.regopy as rego

policy = """
package multicloud.authz

default allow = false

allow {
    input.method == "GET"
    input.path =="/api/data"
    input.user.roles[_] == "data-reader"
    input.user.mfa == true  # Zero Trust requirement
}
"""

def evaluate_access(request):
    result = rego.evaluate(policy, input=request)
    return result['allow']
```

```
# Usage
request = {
    "method": "GET",
    "path": "/api/data",
    "user": {"roles": ["data-reader"], "mfa": True}
}
print(evaluate_access(request))   # Returns True/False
```

Key Strategies to Use

To strengthen ZTA, various strategies can be used to improve access management and support long-term security needs, as described here:

a. **Identity-centric security:** This is a model that ensures every approach has the right categorization. Every user has to indicate their identity before getting onto the platforms. This can be done through multifactor authentication, ensuring mandatory identity for the platforms. Additionally, working with continuous adaptive risk and trust assessment can ensure access is adjusted based on risks that have been detected. These aspects make it easier to manage identities.

b. **Use of least privilege:** Addressing privileges using just-in-time access can help to ensure there are no instances of standing privileges on the platform. More to the point, ABAC offers a better way to handle permissions and adapt to contexts.

c. **AI-driven threat detection:** This is an intervention that can help in handling logs analysis to detect real-time anomalies, leading to better management and handling. More to the point, using GCP SCC and AWS GuardDuty and Azure Defender provide a chance to analyze these logs, detect real-time challenges, and work toward establishing beneficial steps to advance valuable results in the organization.

d. **Using end-to-end encryption and secure access:** The configuration of mutual TLS will help create better steps to manage service-to-service communication. Using a software-defined perimeter helps to hide resources until their process of authentication has been successful in the organization.

Zero trust is increasingly being applied in hybrid cloud environments. The use of zero trust ensures passwordless authentication by working with FIDO2 and biometrics. Additional advances in confidential computing by encrypting data in use helps to mark engagement of ZTA to offer a new layer of security. Increasingly, working with Autonomous Policy Enforcement helps to create the best channel of making decisions by AI, ensuring that the protection is pushed a notch higher for these organizations. Thus, the use of the zero trust improves security for hybrid cloud environments.

References

Ahmadi, S. (2024). Zero trust architecture in cloud networks: Application, challenges and future opportunities. *Journal of Engineering Research and Reports*, *26*(2), 215-228.

Hong, J., Dreibholz, T., Schenkel, J. A., & Hu, J. A. (2019). An overview of multi-cloud computing. In *Web, Artificial Intelligence and Network Applications: Proceedings of the Workshops of the 33rd International Conference on Advanced Information Networking and Applications (WAINA-2019) 33* (pp. 1055-1068). Springer International Publishing.

Ike, J. E., Kessie, J. D., Okaro, H. E., Ezeife, E., & Onibokun, T. (2025). Identity and Access Management in Cloud Storage: A Comprehensive Guide.

Imran, H. A., Latif, U., Ikram, A. A., Ehsan, M., Ikram, A. J., Khan, W. A., & Wazir, S. (2020, November). Multi-cloud: a comprehensive review. In *2020 ieee 23rd international multitopic conference (inmic)* (pp. 1-5). IEEE.

Julakanti, S. R., Sattiraju, N. S. K., & Julakanti, R. (2022). Multi-Cloud Security: Strategies for Managing Hybrid Environments. *NeuroQuantology*, *20*(11), 10063-10074.

Khambam, S. K. R., & Kaluvakuri, V. P. K. (2023). Multi-Cloud IAM Strategies For Fleet Management: Ensuring Data Security Across Platforms.

Patel, S. (2024). CLOUD SECURITY BEST PRACTICES: PROTECTING YOUR DATA IN A MULTI-CLOUD ENVIRONMENT.

Ramalingam, C., & Mohan, P. (2021). Addressing semantics standards for cloud portability and interoperability in multi cloud environment. *Symmetry*, *13*(2), 317.

Sidharth, S. (2021). Multi-Cloud Environments: Reducing Security Risks in Distributed Architectures.

Somanathan, S. (2023). Securing the Cloud: Project Management Approaches to Cloud Security in Multi-Cloud Environments. *International Journal of Applied Engineering & Technology*, *5*(2).

Spinella, E. F. (2025). Kubernetes Workload Identity Federation.

Vasanthi, G. (2024). Cloud Migration Strategies for Mainframe Modernization: A Comparative Study of AWS, Azure, and GCP.

Yeboah-Ofori, A., Jafar, A., Abisogun, T., Hilton, I., Oseni, W., & Musa, A. (2024, August). Data Security and Governance in Multi-Cloud Computing Environment. In *2024 11th International Conference on Future Internet of Things and Cloud (FiCloud)* (pp. 215-222). IEEE.

Zhao, H., Benomar, Z., Pfandzelter, T., & Georgantas, N. (2022, December). Supporting multi-cloud in serverless computing. In *2022 IEEE/ACM 15th International Conference on Utility and Cloud Computing (UCC)* (pp. 285-290). IEEE.

CHAPTER 9

Securing Sensitive Data with IAM

I. Managing Access to BigQuery, Cloud Storage, and Spanner

Identity and Access Management (IAM) is key to enhancing security in GCP architecture. The use of IAM enables an administrator to provide access to individual accounts, ensuring that they can use specific resources to help them create valuable solutions at all times. Creating the best configurations helps enable authorized users and services on the network to make sure the platform is secure. Thus, using a cloud platform is beneficial path to advancing an organization's needs on the GCP platform. Services such as Cloud Storage, BigQuery, and Spanner have to be secured to improve data management and modeling. Thus, the security of GCP has to consider these services to advance the safety of a system.

Fundamentals in IAM for BigQuery, Cloud Storage, and Spanner

There are some fundamental issues to be aware of when managing these GCP services, such as understanding the IAM roles and permissions. You need to categorize the roles. For example, primitive roles are broad and provide execution-based options such as *owner, viewer*, and *editor*. Additionally, predefined roles depend on an individual service and enhance the capacity to work within it (Cabianca, 2024). The service-specific roles include elements such as *roles/bigquery.dataviewer*. There are also provisions for custom roles, which can be used to meet unique requirements. Custom and predefined roles are critical when managing sensitive data because they offer individualized control of the data. This works better than the primitive roles for the GCP architecture.

Other elements are the resource hierarchy and inheritance. The GCP resources have to inherit IAM policies from different parent resources. Each of these elements provides an incremental level of support and ways to add value to the IAM management. Therefore, the policies have to be managed in ways that consider the critical components, leading to a better and more definitive way to handle policies from different angles. For instance, the use of the resource-level policies will help override inherited ones, leading to a better level of securing resources and much better creation for the GCP resource.

Handling Access to BigQuery

BigQuery is key in any GCP architecture, since it serves as storage for structured data, financial records, PII, and business intelligence information. You have to securely manage BigQuery and can use different roles to do so, , as summarized in Table 9-1 (see also Figure 9-1).

Table 9-1. Roles in BigQuery

Role	Description	Use Case
Roles/bigquery.admin	Exercises full control over available datasets.	Administrators in an organization have to regulate data in various resources.
Roles/bigquery.dataEditor	Have the rights to edit data, but not editing schema on the platform.	They can be used by data engineers who have to modify tables and ensure every piece of information checks out.
Roles/bigquery.jobUser	They have the right to run queries, but they do not have any direct access to data.	Different apps and platforms that have to execute queries, leading to a remarkable scope and potential of advancing solutions to pertinent issues.
Roles/bigquery.dataViewer	They have the rights to read table metadata and data.	Permissions granted to analysts on the platform to ensure that they have to gain access to query and interpret whatever information is articulated.

CHAPTER 9 SECURING SENSITIVE DATA WITH IAM

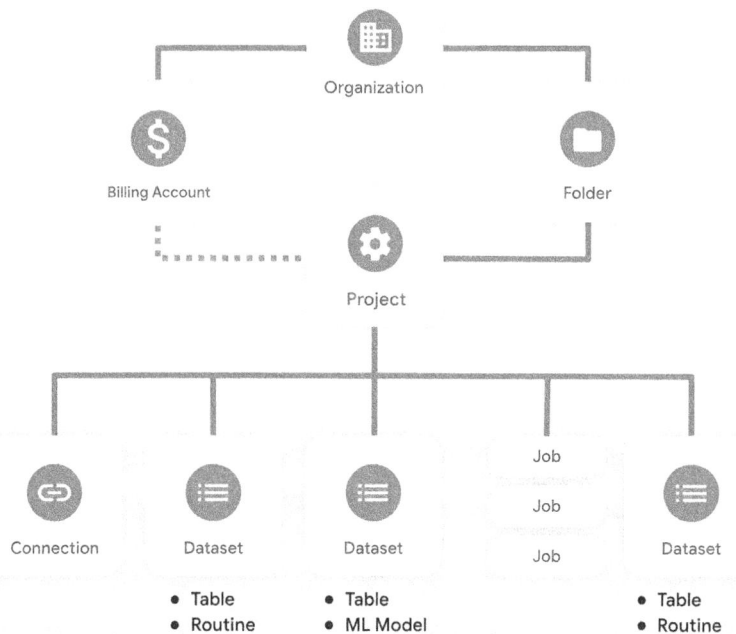

Figure 9-1. *GCP BigQuery roles*

Considerably, when working with IAM on BigQuery, different practices have to be considered, with each instance marking the development of a new scope of management for the information (Levadoski et al., 2024). IAM can enhance security and safety for BigQuery, leading to better constructed solutions. These practices include the following:

    a. **Dataset-level ACLS:** This creates a capacity to restrict access within datasets, ensuring dataset-level restriction, instead of project-wide restrictions. This can be conducted through the following code:

```
gcloud projects add-iam-policy-binding PROJECT_ID \
  --member=user:analyst@example.com \
  --role=roles/bigquery.dataViewer \
  --condition='expression=resource.name.startsWith("projects/PROJECT_ID/
    datasets/sensitive_dataset")'
```

133

b. **The use of column-level security:** BigQuery can implement column-level ACLs or data masking activities. This can be conducted by working with the role *roles/bigquery.filteredDataViewer*. The entire approach enables an increased level of security, leading to better column-level security provision.

c. **Row-level security (RLS):** This can be conducted by ensuring provision of row access policies, each restricting access for certain users based on their attributes on the system. This enables support for whatever actions have to be partaken on the system.

d. **Conditional access:** In certain instances, conditional access has to be established, ensuring that the BigQuery service works within limited timelines or IP ranges. This conditional access capability offers the potential to craft solutions in every detailed IAM use for the GCP resource. The following code can be used for conditional access:

```
gcloud projects add-iam-policy-binding PROJECT_ID \
  --member=user:contractor@example.com \
  --role=roles/bigquery.dataViewer \
  --condition='expression=request.time.getHours("America/New_York") >= 9 &&
    request.time.getHours("America/New_York") <= 17'
```

Handling Access to Cloud Storage

Cloud storage is a key resource on the GCP cloud service. The cloud storage is a database for unstructured data. The platform can be used to handle backups, media, and logs, ensuring that they have the right categorization of information (Roy et al., 2021). The platform therefore demands a strict access control, each working to enhance and achieve optimal results. Major roles that can be used on the Cloud Storage service are summarized in Table 9-2.

Table 9-2. Roles in Cloud Storage

Role	Description	Use Case
Roles/storage.objectAdmin	Have the full capacity to read/write and delete an object.	This can be used by DevOps teams to achieve a desired response.
Roles/storage.objectViewer	Have the right to read objects.	This is applied in reporting tools that are available on the system.
Roles/storage.admin	Overall control over the bucket.	Administrators having the capacity to view, look into, and handle the objects at different levels.

There are different practices that have to be applied to achieve stellar IAM management on Cloud Storage. These practices include aspects such as the following:

i. **Bucket-level IAM policies:** These policies enable the provision of roles at the bucket level. The approach is beneficial for individual access provision, rather than having a wide and nondefined access provision model. This can be done using the following code:

```
gcloud storage buckets add-iam-policy-binding gs://sensitive-bucket \
  --member=serviceAccount:etl-sa@PROJECT_ID.iam.gserviceaccount.com \
  --role=roles/storage.objectAdmin
```

ii. **The use of object-level ACLs:** The object-level ACLs have to be used sparingly, even though they have a high level of efficiency in advancing solutions to a critical end (Cemeh, 2024). They have the capacity to provide granular file-level access, as detailed in Figure 9-2, leading to a much better construction of whatever security approaches have to be developed.

CHAPTER 9   SECURING SENSITIVE DATA WITH IAM

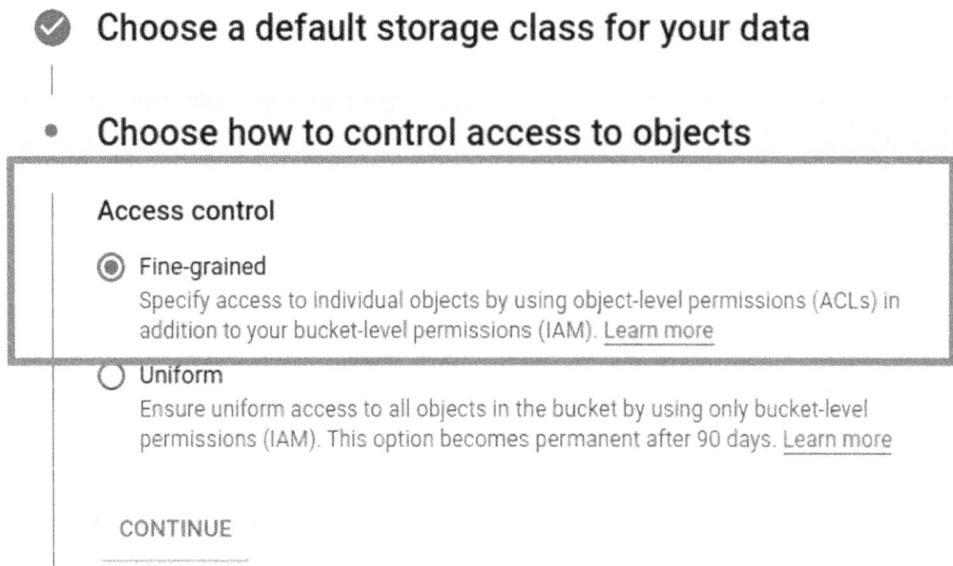

Figure 9-2. GCP Cloud Storage object-level ACLs

    iii. **Signed URLs and signed policies:** Time-limited access can be provided via signed URLs to grant temporary access. The temporary access leads to better security.

    iv. **Conditional access:** Conditional access can ensure the provision of access only within a specific VPC, leading to a much better angle of considering selection needs of the Cloud Storage. The conditional access can be enacted using the following code:

```
gcloud storage buckets add-iam-policy-binding gs://secure-bucket \
  --member=user:remote-worker@example.com \
  --role=roles/storage.objectViewer \
  --condition='expression=request.network == "projects/PROJECT_ID/global/
  networks/secure-vpc"'
```

Handling Access to Cloud Spanner

Cloud Spanner is a resource on GCP that acts as a relational database to manage the cloud functionalities (Kingsley, 2023). Different roles can be designated on Cloud

Spanner, ensuring that every segment works well and enhancing functionality and appropriate outcomes at all times. The roles are summarized in Table 9-3.

Table 9-3. Roles in Cloud Spanner

Role	Description	Use Instance
Roles/spanner.databaseUser	Has the capacity to read and write data based on the capacity that they have been offered.	Application Services.
Roles/spanner.databaseReader	Has the capacity to read data and interpret various ways that it can be used on the system.	Can be used in reporting applications, ensuring tabulation and provision of accurate ways to depict and detail information as needed.
Roles/spanner.admin	Provision of Full Control on the platform.	This is provided for database administrators, ensuring that they can manage every viable activity conducted on the platform.

The following practices can enhance the use of Cloud Spanner:

a. **Instance level vs. database-level roles:** This ensures is a specific level of access to the users. Users can only access data at the database level, and there are critical measures for securing the nonessential data elements (Shashi, 2023).

b. **Federated access via IAM:** This ensures that service accounts can be used, which increases safety, instead of user-based access.

c. **Fine-grained access management using database roles:** This is an advancement that ensures the use of tools such as PostgreSQL-style roles to provision column-level restrictions as desired.

d. **Conditional IAM regulations:** These ensure that only a specific database can hold edit or ownership roles. This can be done using the following code:

```
gcloud spanner databases add-iam-policy-binding DB_NAME \
  --instance=INSTANCE_NAME \
  --member=user:dev@example.com \
```

```
--role=roles/spanner.databaseUser \
--condition='expression=resource.name.endsWith("/databases/prod_db")'
```

Auditing and Monitoring Access

The process of auditing and monitoring access for these GCP services demands critical skills to achieve remarkable results. Cloud Audit Logs can ensure that there is access to data from the BigQuery, Spanner, and Cloud Storage platforms. In essence, these services offer lots of tools for monitoring the admin activity and looking at the data. The events that they can investigate include Data Write/Read events, which results in better insights into the information used on the platform (Mulder, 2024). Setting alerts on anomalies will also give insights into unusual access patterns. Moreover, setting conditional IAM regulations will ensure that access can be monitored, including attempts at accessing the cloud platforms and services wherever they are located.

II. Using Customer-Managed Encryption Keys (CMEK) for Data Security in Google Cloud

Encryption is a major step to safeguarding data within cloud platforms. GCP offers default encryption capacities, which has the potential to manage data both at rest and data in transit, as indicated in Figure 9-3. Notably, organizations have to work with confidential information and highly regulated information and thus need to continually ensure safety in any activity they partake on the data (Joshi & Phadke, 2024). To increase the security management, organizations have to use additional encryption approaches such as customer-managed encryption keys (CMEKs) to implement advanced data protection models.

CMEK is an approach where an organization manages their encryption keys instead of working with the default encryption provided by GCP. CMEK ensures that an enterprise can generate and handle storage and enable access of the encrypted files using the Cloud Key Management Service (KMS).

CMEK has a stark difference with Google Managed Keys, ensuring an increasingly essential point of managing the keys. Google Managed Keys are a default encryption type that is automatically generated and managed by the GCP platform. It provides basic encryption that the customer has no control over. Meanwhile, CMEK are keys that are wholly generated and managed by the customer through the Cloud KMS. CMEK enables

CHAPTER 9 SECURING SENSITIVE DATA WITH IAM

customers to handle key rotation, audit logging, and access control activities (Moore et al., n.d). The approach also gives benefits when managing compliance to various provisions such as GDPR and HIPAA. Using CMEK therefore provides major benefits like ensuring that organizations have complete ownership over the encryption keys that they generate and can also provide them with more control. This allows them to be disabled and rotated when they are being applied. In addition, ensuring regulatory compliance within CMEK is easier, and CMEK offers a fine-grained access management approach, ensuring that the IAM policies can determine and even restrict whoever can manage the keys on the platform. Cloud Audit can be used alongside CMEK, making it simpler to audit and locate the access points of every key.

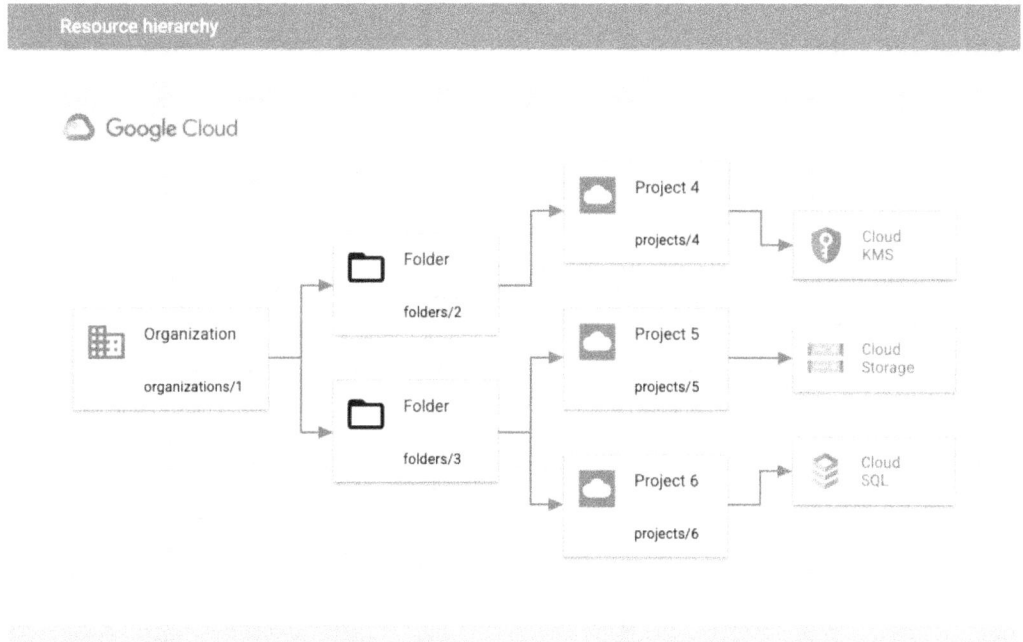

Figure 9-3. *CMEK use in GCP*

Implementing CMEK Across Key GCP Services

CMEK has to be implemented correctly on various GCP resources and services. To implement CMEK for Google Cloud Storage, authorization has to be used to ensure only the correct users can access the platform (Guptha & Murali, 2021). This approach can help organizations conduct data backups and enables a secure document storage mechanism, leading to better development techniques. The implementation can be bolstered with strict key access to necessary roles, monitoring the key usage and working with CMEK at bucket creation, to ensure a consistency in encryption activities.

When working with BigQuery, CMEK can encrypt datasets and temporary files. This ensures that the use of analytics on sensitive data such as financial records and PII can be curtailed, as well as working with multi-tenant data isolation approaches. Thus, key generation and management can assist in creating the desired results (Carvalho et al., 2019). To improve CMEK usage with BigQuery, datasets should be created with CMEK from the start. This allows for better key separation across development, production, and during key rotations, helping reduce the risk of key exposure at all times.

CMEK can also be used for Cloud Spanner, ensuring an additional layer of security and better transactional data management advances. Using CMEK is a chance to work with strict access control and regulatory compliance needs to ensure good information management techniques. CMEK can be used at instance creation to achieve the desired security level. CMEK also works to ensure high availability of storing keys in different regions. Thus, CMEK achieves the best outcome possible.

Best Practices for Managing CMEK

Managing CMEK within GCP is critical to achieve success. The following are CMEK best practices:

a. **Centralized key management with KMS:** To enhance the functionality of the CMEK approach, keys have to be organized hierarchically so they can be identified and marked for deployment. Key rings have to be deployed so that group-related keys can be identified easily.

b. **Lifecycle management and key rotation interventions:** Keys have to be rotated from one period to the next. Additionally, key versions have to be maintained, ensuring that legacy data can be decrypted to better identify and manage the keys at all times.

c. **Strict IAM control for the keys:** Following the policy of least privilege will ensure that there is a considerable access only to the desired policies and requirements of the IAM permissions. Nonetheless, permissions should not be granted broadly.

d. **Disaster recovery planning:** Enterprises have to implement disaster recovery planning. They have to ensure they back up key materials and cannot get access when disasters occur. The key destruction approaches also have to be defined and documented.

e. **Monitoring, auditing, and alerting:** The CMEK setup should support detailed audit logging to track KMS operations clearly. Alerts should also be in place to detect suspicious activity and respond quickly, helping maintain the platform's security at all times.

CMEK provides enterprises with a great capacity to control, ensure compliance adherence to, and even secure their keys. The use of CMEK works better compared to the Google Cloud data encryption mechanisms. Thus, using cloud KMS to leverage the CMEK functionalities increase the capacity and chance to continually manage the CMEK platform to achieve the desired results (Estrin, 2022). More to the point, CMEK gives clients more responsibility when managing encryption compared to GCP's default encryption, which is handled by the cloud provider. Using CMEK with key lifecycle management allows companies to maintain stronger control and implement more advanced key management practices.

III. Fine-Grained Access Control for Sensitive Resources

Sensitive data within GCP architecture has to be appropriately handled, ensuring no data breaches on these platforms. The use of personally identifiable information, intellectual property, and financial records has to be safeguarded, ensuring an appropriate level of access and reporting on these resources, leading to lesser liability on these platforms.

Fine-grained access control is a great way to handle excessive privileges and data leaks. The access control model creates minimal attack surfaces, as the access is provided for only the necessary functions and services. The access complies with regulations, providing a framework for handling data and aligning it with desired outcomes (Murthy et al., 2024). Enforcing the least privilege policy prevents insider threats, making breaches less common.

Fine-grained access can be handled using different approaches. For example, attribute-based access control is key to ensuring that permissions are provided based on resource metadata and user attributes. Environmental conditions can also be used to ensure that the access is well managed and that functionalities are implemented in a way that models the safety and security of resources needed.

More to the point, column-level and row-level security can be used to handle fine-grained access in Big Query. In this case, the sensible management of data through policy tags in BigQuery is a chance to improve safety. For example, row-level security filters rows based on user identity and group membership features.

Resource-level IAM conditions can also be used to ensure fine-grained access control. Using context-aware restrictions can ensure permissions are provided based on major elements such as temporal constraints, compliance of devices from which requests are sent, and IP restrictions. These aspects will ensure that access is managed with regard to critical issues.

Organizations can also provide guardrails in their policies to help with the management of the fine-grained access controls. Using VM restrictions for external IP assignments is a chance to consistently provide value. Using right policies will ensure you improve organizational safety.

Best Practices for the Implementation of Fine-Grained Access

Fine-grained access can be handled by following certain best practices:

  i. **Policy of least privilege:** User and service accounts have to be provided with the least possible privileges that they need to help advance their needs.

  ii. **Custom IAM roles:** Organizations have to provide custom roles to give them an edge in addressing major needs (Albulayi et al., 2020).

  iii. **Policy intelligence tools:** Tools like IAM Recommender help identify the most appropriate policies for each situation.

iv. **Auditing and monitoring:** Enterprises have to monitor, evaluate, and provide alerts for anomalies on their system. This approach supports a step-by-step improvement in the efficiency of fine-grained access controls.

Fine-grained access controls can ensure sensitive data is secure on the GCP architecture. Working with features such as row/column-level security, ABAC, VPC service controls, and IAM conditions, it becomes much simpler to have context-aware permissions and reach an increased level of compliance when managing and sustaining organizational features.

IV. Industry-Specific Use Case: Protecting Healthcare Records with CMEK and IAM

The healthcare sector has to work with patient information and comply to requirements in HITRUST, GDPR, and HIPAA.

CMEK and IAM help improve compliance by ensuring that only authorized users can access patient data in Cloud Storage, the Healthcare API, and BigQuery. CMEK encryption, combined with IAM, supports role- and attribute-based access, making sure the right people can access information when needed (Sun et al., 2021). On the other hand, fine-grained IAM policies can restrict access through RBAC and ABAC, and categorizing the roles helps to designate duties and activities that can be conducted across platforms, as indicated in Figure 9-4. These advancements provide an effective way to manage and improve information handling. They also enable audit logging, which helps track access and changes to data. Overall, these tools are essential for consistently managing healthcare information securely and efficiently.

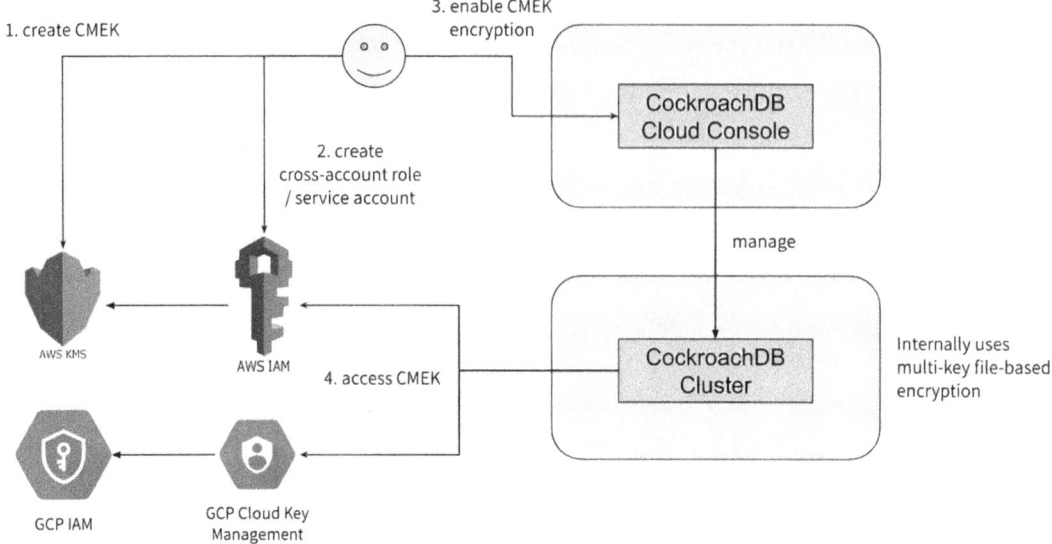

Figure 9-4. Healthcare records using IAM and CMEK

References

Albulayhi, K., Abuhussein, A., Alsubaei, F., & Sheldon, F. T. (2020, January). Fine-grained access control in the era of cloud computing: An analytical review. In *2020 10th Annual Computing and Communication Workshop and Conference (CCWC)* (pp. 0748-0755). IEEE.

Cabianca, D. (2024). Configuring Access. In *Google Cloud Platform (GCP) Professional Cloud Security Engineer Certification Companion: Learn and Apply Security Design Concepts to Ace the Exam* (pp. 15-175). Berkeley, CA: Apress.

Carvalho, D., Morais, J., Almeida, J., Martins, P., Quental, C., & Caldeira, F. (2019, July). A technical overview on the usage of cloud encryption services. In *European Conference on Cyber Warfare and Security* (pp. 733-XI). Academic Conferences International Limited.

Estrin, E. (2022). *Cloud Security Handbook: Find out how to effectively secure cloud environments using AWS, Azure, and GCP*. Packt Publishing Ltd.

Guptha, A., & Murali, H. (2021, May). A comparative analysis of security services in major cloud service providers. In *2021 5th International Conference on Intelligent Computing and Control Systems (ICICCS)* (pp. 129-136). IEEE.

Joshi, H., & Phadke, N. (2024). Cryptographic Bastions: Mastering Cloud Security through Advanced Access Control and Encryption Strategies. In *Cloud Security* (pp. 76-100). Chapman and Hall/CRC.

Kingsley, M. S. (2023). Cloud Platform. In *Cloud Technologies and Services: Theoretical Concepts and Practical Applications* (pp. 143-156). Cham: Springer International Publishing.

Levandoski, J., Casto, G., Deng, M., Desai, R., Edara, P., Hottelier, T., ... & Volobuev, Y. (2024, June). BigLake: BigQuery's Evolution toward a Multi-Cloud Lakehouse. In *Companion of the 2024 International Conference on Management of Data* (pp. 334-346).

Moore, T. L., Conlon, S. S., Hewarathna, A. U., & Mailewa, A. B. Encryption Methods and Key Management Services for Secure Cloud Computing: A Review.

Mulder, J. (2024). *Multi-Cloud Administration Guide: Manage and Optimize Cloud Resources Across Azure, AWS, GCP, and Alibaba Cloud*. Walter de Gruyter GmbH & Co KG.

Murthy, J. S., Siddesh, G. M., & Srinivasa, K. G. (Eds.). (2024). *Cloud Security: Concepts, Applications and Practices*. CRC Press.

Roy, A., Banerjee, A., & Bhardwaj, N. (2021). A study on google cloud platform (gcp) and its security. *Machine Learning Techniques and Analytics for Cloud Security*, 313-338.

Shashi, A. (2023). Designing Applications for Google Cloud Platform. *Designing Applications for Google Cloud Platform*.

Sun, J., Yuan, Y., Tang, M., Cheng, X., Nie, X., & Aftab, M. U. (2021). Privacy-preserving bilateral fine-grained access control for cloud-enabled industrial IoT healthcare. *IEEE Transactions on Industrial Informatics*, 18(9), 6483-6493.

Семен, М. Л. (2024). Comparative analysis of security models in cloud platforms. *Известия Томского политехнического университета. Промышленная кибернетика*, 2(2), 1-16.

CHAPTER 10

AI-Driven Identity and Access Management

I. Introduction to AI in IAM

Identity and Access Management (IAM) is crucial in cloud security. It therefore needs to adjust to the enterprise security needs of an organization. Advances in AI have boosted security across many areas, and today's businesses use AI in IAM to keep their cloud systems safe.

AI in Cloud Security

AI has the capacity to analyze vast datasets, detect anomalies, analyze information, and predict risks. AI can also enforce policies. Nonetheless, using AI in IAM can find access patterns, gain threat intelligence, and get an overview of user behavior. You can use AI cloud security management to detect anomalies, consider access requests, find behavioral patterns, and detect suspicious activities before they affect the overall functionality of the system. AI works by enabling access prediction mechanisms, as indicated in Figure 10-1. This model manages ML mechanisms to control access by reviewing excessive permissions, historical data, and the policy of least privilege for the best results (Mamidi, 2024). Using automated policy recommendations helps identify existing policies and improve them, adding value to AI-powered systems. This shows how AI in cloud security supports continuous improvements at every management level.

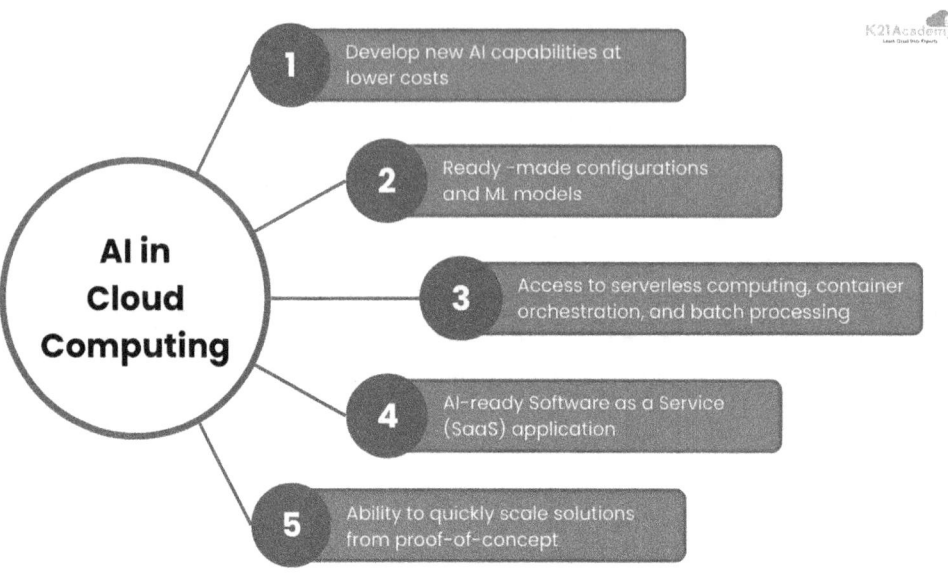

Figure 10-1. AI in cloud security

Benefits of Using AI in IAM

Using AI in IAM encourages better performance and reinforces security at all levels. Primarily, AI helps with the proactive nature of threat detection. It ensures that the AI-powered IAM system can flag anomalies to help prevent breaches before they occur. This is a major security challenge without AI (Rehan, 2024).

Using adaptive access control helps model risks as they occur within the enterprise system. Notably, the use of the adaptive access control enables several mechanisms to achieve the management needs. The use of MFA, for example, ensures that any risky access attempts can be flagged, resulting in greater security for the system.

Finally, using AI in IAM reduces the administrative overhead. AI leads to automation, increases accuracy, and improves latency. Working with AI saves time when handling projects and reduces administrative burden, leading to a better development for the organization when protecting their data.

AI-Driven IAM Tools in GCP

GCP offers an interactive platform that integrates both AI and ML tools in the IAM system. AI and ML tools can enhance threat detection, identity verification, and policy optimization. For example, Policy Intelligence helps GCP to ensure that AI analyzes the IAM policies and provides recommendations on the configurations that work well. Policy Intelligence can optimize on safety, minimize security risks, and lead to a remarkable level of safety on the platform (Stutz et al., 2024).

GCP also uses Cloud Armor to help detect and mitigate malicious traffic patterns as they occur within the cloud network. Cloud Armor acts as a DDoS and web application firewall, with an adaptive protection mechanism, using ML models to handle traffic, distinguishing malicious from legitimate requests, and blocking attacks before they become a full scale.

GCP also provides AI-enabled APIs to help manage the identity verification process. For example, Vision AI analyzes identity documents to enhance fraud detection. Meanwhile, Recommendation AI makes suggestions for the policies, as it looks into usage trends and imminent issues within the system. The Natural Language API works on user behavior, looking into social engineering attempts, conducting text analysis, and detecting phishing attempts on the platform. Each of these AI-enabled APIs work within the GCP framework to increase safety at all levels of implementation.

Considerably, AI has helped to advance IAM, ensuring a more adaptive, efficient, and intelligent IAM system. The use of anomaly detection, automated policy management, and predictive access control has ensured that GCP can increase IAM efficiency, leading to better security for the organization.

II. Anomaly Detection with AI

Anomaly detection, when powered with AI, adds increasing value to an organization. By monitoring network activity and identifying unusual patterns, AI helps detect potential breaches and supports critical operations.

Machine learning (ML) enables this by establishing a baseline of normal system behavior. Actions outside this baseline are flagged as anomalies, which can then be investigated to improve IAM systems over time (Ait Chikh, 2023). Therefore, deviating from any of these actions are considered as a case for alert on the system. The main indicators of an anomaly within the system include aspects such as the following:

i. Excessive permission requests

ii. Unusual login times or locations

iii. Abnormal API calls

Anomaly detections on IAM can be handled using the following Python code:

```python
from google.cloud import logging_v2
from google.cloud import recommender_v1

client = logging_v2.LoggingServiceV2Client()
parent = "projects/YOUR_PROJECT_ID"

# Fetch unusual IAM grants from Cloud Audit Logs
filter_ = """
logName="projects/YOUR_PROJECT_ID/logs/cloudaudit.googleapis.com%2Factivity"
protoPayload.methodName="SetIamPolicy"
severity>=WARNING
"""

entries = client.list_log_entries(request={"resource_names": [parent], "filter": filter_})

# Get IAM recommender suggestions
recommender_client = recommender_v1.RecommenderClient()
recommendations = recommender_client.list_recommendations(
    parent="projects/YOUR_PROJECT_ID/locations/global/recommenders/google.iam.policy.Recommender"
)
```

Using Policy Troubleshooter and Cloud Audit Logs with AI

GCP provides a series of AI-powered tools that can help detect anomalies. Policy Troubleshooter identifies the reason why a service account or user was given access to a particular resource. This reveals any instances of misconfiguration or overly permissive roles within the system. Cloud Audit Logs ensures a record for every access event, which provides the opportunity to detect suspicious patterns within the system. To enhance the safety of the system, security teams have to carry out different activities that work in combination to craft the best outcome on the system, as indicated in Figure 10-2. Notably, security teams can use both Policy Troubleshooter and Cloud Audit Logs to identity policy violations, trace any instance of unauthorized access, and automate remediation through the use of AI-generated recommendations.

Financial companies offer use AI to detect threats. Most financial companies utilize a hybrid cloud infrastructure, ensuring that they have access patterns from employee accounts. The data access patterns have to be recorded by the Cloud Audit Logs, which reveals any instance of violation. Examples of violations include downloading large volumes of sensitive records and accessing customer databases during nonworking hours (Saminathan, 2024). This issues can then be investigated with the Policy Troubleshooter platform to reveal more information about whatever categories and needs have to be addressed. Policy Troubleshooter helps identify outdated role assignments that could allow unauthorized access. Using this tool with other GCP services can improve security in financial organizations.

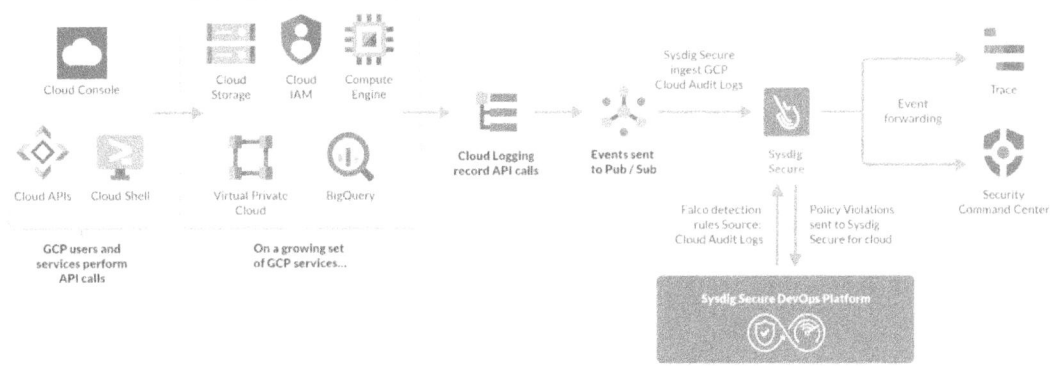

Figure 10-2. Using Cloud Audit Logs

III. Automating Policy Management with AI

Using IAM policies within an enterprise is a complex affair, requiring companies to constantly be improving their functionality. The traditional policy management approach often leads to errors that result in security gaps, excessive permissions, and compliance risks. Automation through AI provides a better way to manage the IAM needs, ensuring increased proficiency in access controls and enforcing the policy of least privilege.

AI-Based Policy Recommendations in GCP

GCP works with Policy Intelligence to analyze and optimize IAM policies. The service evaluates existing policies and permissions, locates the over-privileged accounts, and suggests changes to achieve the same results with the least policies provided to the users. Policy Intelligence automated and advanced role analysis, scanning IAM roles and locating the right ways to produce output, but with minimal privilege assigned to the resources.

More to the point, a policy simulator assists in locating and working with policy changes before they are rolled out to the entire IAM system (Roy et al., 2021). The simulator uses a controlled environment to test recommended polices, determine their impact, and address the organizational needs. Therefore, the simulator makes it much simpler to categorize, handle, and manage the AI-based recommendations for the policies. AI also works with security benchmarks such as CIS and NIST to ensure remediation. It flags deviations and strives for the accurate and proper management of the IAM system.

The policy recommendations can be implemented in the code as follows:

```
from google.cloud import recommender_v1
from google.cloud import asset_v1

def apply_iam_recommendation(recommendation_name):
    recommender_client = recommender_v1.RecommenderClient()
    request = recommender_v1.MarkRecommendationClaimedRequest(
        name=recommendation_name,
        etag=recommendation.etag
    )
    response = recommender_client.mark_recommendation_claimed(request=request)
    return response

# Get all IAM recommendations
asset_client = asset_v1.AssetServiceClient()
recommendations = asset_client.analyze_iam_policy(
    request={
        "analysis_query": {
            "scope": f"projects/YOUR_PROJECT_ID",
```

```
        "identity_selector": {
            "identity": "user:example@domain.com"
        }
    }
}
)
```

Predicting Least Privilege Access Using Historical Data

AI also enables analysis of historical access patterns to predict the minimum permissions that any user needs to ensure they can fulfill their designated roles. The Recommender API on GCP uses ML to generate insights that will provide better management of the organization's needs and recognize trends for each user on the platform. Notably, the approach offers permission management and help reduce the attack surface. It can enhance historical data to result in better management of IAM.

Real-World Example: Automating Role Assignments in a Multi Project Organization

A multinational organization works on several GCP projects, an aspect that has led to several inconsistencies in role management and assignment and that has let to security vulnerabilities in the system because of how they handle employee onboarding and role changes within the organization. The organization worked to implement an AI-enabled policy automation approach, which helped them to address each of the challenges that they faced (Olobanji et al., 2024).

Automated onboarding was essential for the organization, as it introduced a new system for managing role assignments. New data analyst hires were automatically given basic access to BigQuery, allowing them to work with datasets from day one, as shown in Figure 10-3. AI was then used to monitor their activity over a 30-day period to ensure they could perform their tasks effectively. Based on this monitoring, AI also recommended additional permissions they might need, as well as policies or access rights that should be revoked to maintain security and ensure they only had access to what was necessary.

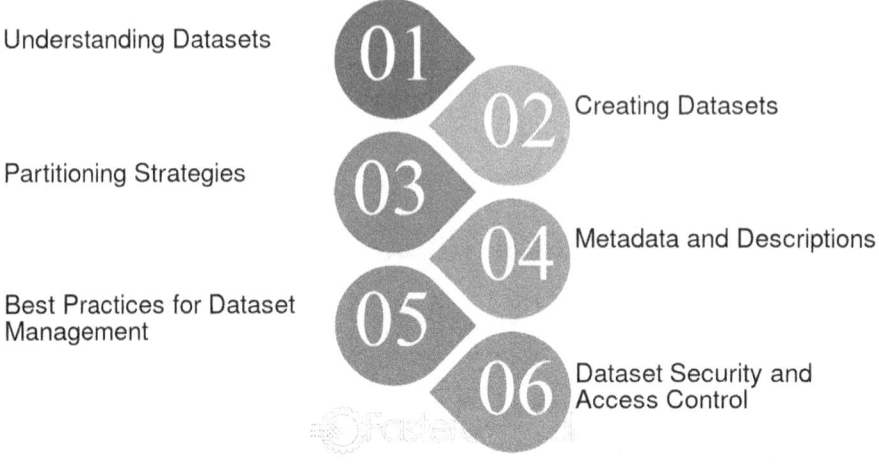

Figure 10-3. BigQuery and handling of data sets

More to the point, the dynamic role adjustments were conducted using AI, ensuring that any outdated permissions were handled in the right manner. They implemented automatic revocation for temporary access to sensitive services, ensuring that the provided timeline was not exceeded (Kumar & Sharma, 2025). The AI-based approach resulted in an automated audit, ensuring that there were monthly access reviews that included looking at flagged anomalies and patterns (Ike et al., 2025). Thus, the use of the AI in the multinational organization helped to reduce excessive privileges and permissions in the organization with a 40% margin.

Implementing AI-Driven Policy Management

AI-based policy automation activities can be conducted on GCP to ensure increased efficiency. However, the steps needed to carry out AI automation should begin with enabling Policy Intelligence on the GCP console and ensuring that Recommender is activated to assist in locating and mapping whatever actions are needed on the platform.

On the other hand, AI suggestions have to be reviewed to ensure the best outcome and to handle meaningful appeals (Rimon et al., 2024). These suggestions will help organizations handle their policy management. Notably, additional steps have to be

CHAPTER 10 AI-DRIVEN IDENTITY AND ACCESS MANAGEMENT

used to monitor and refine the recommendations. Cloud Audit Logs can track policy changes, administer their needs, and achieve sustainable results in whatever capacities are documented. Thus, using AI to this level will enhance the IAM functionality, creating better suggestions for permissions and role management needs.

IV. AI-Powered Identity Verification

With the sophistication of modern-day cyber threats, traditional identity verification approaches have proven insufficient. Integrating AI with identity verification creates a advancement toward ensuring better security and catering the progressive nature of the cloud computing needs. The use of tools on GCP such as Google's Vision AI can automatically perform identity checks, creating stronger MFA systems and fraud detection models.

Leveraging AI Tools for Biometric Verification

AI-driven biometric verification employs fingerprint scanning, facial recognition, and document analysis to analyze with high precision. Google Vision AI enhances this by bringing in better facial recognition, comparing live selfies with government-issued identification documents to certify identities. Furthermore, authentication of documents is made easy using Google Vision, as it finds documents that might have been tampered with or forged, checks watermarks and holograms, and analyzes text. Liveness detection confirms the user is physically present and not using deepfakes or photos. It does this by analyzing small movements, helping improve system security (Banu, 2024). Using these mechanisms provides can reduce impersonation and streamline user onboarding activities for various organizations such as e-commerce applications, healthcare organizations, and banks. The Google Vision AI can be coded as follows to help with the biometric verification process on the cloud system:

```
from google.cloud import vision_v1

def verify_identity(image_path, id_image_path):
    client = vision_v1.ImageAnnotatorClient()

    # Face detection in selfie
    with open(image_path, "rb") as image_file:
        content = image_file.read()
```

```
    image = vision_v1.Image(content=content)
    face_response = client.face_detection(image=image)

    # ID document text extraction
    with open(id_image_path, "rb") as id_file:
        id_content = id_file.read()
    id_image = vision_v1.Image(content=id_content)
    text_response = client.document_text_detection(image=id_image)

    return {
        "face_detection": face_response,
        "id_text": text_response
    }
```

Automating Identity Verification Workflows

Using manual identity takes a lot of time and money. Manual identity approaches also have a high potential for human error. Integrating AI in the verification process provides greater efficiency as it can extract data from the provided documents. Optical Character Recognition (OCR) can pull details from IDs, cross-checking them with available identity documents in databases.

More to the point, AI can automate risk scoring and assign a likelihood of fraud to different entities based on the anomalies that have been encountered on the system. AI can check for mismatched signatures and inconsistencies in the metadata and behavior of the user accounts (Sugureddy, 2024). AI also provides the benefit of having a continuous learning approach, enhancing accuracy, since it is available to learn over time. This aspect of continuous learning can distinguish legitimate verifications from fraudulent transactions.

Use Case: Implementation of AI-Enhanced Multifactor Authentication

There is a global corporation that operates in seven different countries and needs to ensure that its systems are secure using recent advances in security. The corporation used SMS-based multifactor authentication, but changed to an AI-powered system, which will help them handle SIM swapping and phishing attacks. The AI-based approach included the use of biometric authentication, where employees scan their fingerprints or faces using an application that works alongside Vision AI. Additionally, behavioral analysis is a key factor in the AI-enabled system, ensuring that they can look at the typing speed, login times, and mouse movements to help detect imposters within the system. The AI approach also uses adaptive challenges to add extra security checks whenever it detects any high-risk logins to the platform.

By using an AI-powered MFA system, the multinational company saw a 90% drop in account takeovers, which was a major improvement in account security. It also improved authentication for legitimate users by 50%, helping meet user needs more effectively. Despite these great results, the organization had to continue to manage different issues such as privacy compliance across the multinational branches. Adhering to both GDPR and CCPA within the company is compulsory. Nonetheless, using bias mitigation within the company ensured machine learning could perform equally well on different demographics. This approach also calls for employing explainability, helping to match decisions by the AI.

V. Ethical Considerations in AI-Driven IAM

When using AI in IAM, different ethical issues relating to transparency, bias, and regulatory compliance can affect organizations. These aspects have to be addressed.

Addressing Bias in AI Models for Identity Management

AI bias often comes from the data it's trained on, meaning past biases can affect current results. Common issues include facial recognition errors and geographic bias. To reduce bias, companies should use diverse training data, conduct bias audits, and include human oversight, which can help them adapt to needed changes effectively.

Ensuring Transparency and Fairness in AI-Driven Access Control

AI is a black box, making it difficult for policy makers and users to understand the reasons for its decisions. AI therefore has to be well crafted, enabling better security and management of transparency through use of Explainable AI (XAI), which creates interpretable information in the organization (Robert, 2023). Additionally, having user notifications gives users information on when AI seeks to modify their rights, ensuring there is a chance to appeal. Working with audit logs will also help to maintain transparency, ensuring information exists for use in compliance reviews within the organization.

Aligning AI with Compliance Standards

When cloud computing deals with data, it requires compliance with data handling and management regulations in different jurisdictions. The AI-driven IAM has to comply with these data regulations. Following GDPR's right to explanation means companies have to have a human review of the AI decisions that affect people. It also means organizations should collect only the data they truly need (Anderson, 2022). Nonetheless, the compliance also means working with CCPAs opt-out requirements, making it imperative for users to prevent the use of their data in situations they do not approve. To achieve this, an organization has to ensure privacy by design and work with consent management, ensuring that they disclose information and what will be used in IAM systems.

AI-driven IAM must address ethical risks by reducing bias, ensuring transparency, and meeting compliance standards. This helps businesses use cloud data responsibly and avoid ethical issues.

VI. Emerging Trends and Future Directions

The rapid evolution of AI has shaped IAM and the performance and needs of every enterprise. AI has introduced new capabilities to address traditional inefficiencies. Predictive access controls and AI-enabled zero-trust security both make IAM better in terms of efficiency, security, and agility.

Predictive IAM: Proactive Access Basing on Behavioral Patterns

Predictive IAM uses historical user behavior to look at access needs and automate them within an enterprise. The predictive systems look into the typical behavior that is required in terms of advancing and enabling provisional relevance. The predictive systems look into patterns including aspects like these:

i. **Lifecycle automation:** AI can detect role changes for the employees. This works by handling aspects such as department transfers and promotion activities. The use of an automated system ensures you get updates on access rights based on HR systems so you have the chance to change every activity as needed.

ii. **Role-based activity:** In cases where developers have to access Kubernetes clusters to carry out sprint cycles, the use of AI helps to escalate permissions before the planned deployments take place. This is a key advancement, helping to categorize the roles automatically without having to sign them in every other instance (Duggal & Dave, 2021).

iii. **Contextual signals:** Using ML models ensures the provision of factors such as location, time of day, and device posture, each working to ensure certain logins can be given access to the system and others ones cannot be granted access to the system. Thus, these establishments enable the IAM mechanisms to have a higher level of conditional security even for people signed into the system.

AI-Enhanced Zero-trust Frameworks

Zero-trust frameworks operate on the framework of having a constant demand for verification. The technology marks an easier way to protect data, ensuring that there is an adaptive provision where every user has to work within provided dimensions to achieve the right results. The framework works through the following:

a. **Dynamic policy enforcement:** Instead of using static rules, AI works with real-time access, catering to provisional adjustments at the given time.

b. **Continuous authentication:** Using behavioral biometrics such as mouse movements and keystroke dynamics, AI ensures it provides silent verification to the users so they can flag any instance of a compromised session, leading to a better level of security for the organization (Ajish, 2024).

c. **Threat correlation:** Using AI enables the use of cross references for the IAM logs to ensure that threat intelligence feeds are up to date and to ensure that attackers are blocked even when they impersonate the legitimate users. This works even when there are stolen cookies or the attackers use token replay attacks.

The zero-trust automation approach can be used on BeyondCorp to deploy and manage the automation process. The action can be coded as follows:

```
from google.cloud import accesscontextmanager_v1

def create_zero_trust_policy(project_id):
    client = accesscontextmanager_v1.AccessContextManagerClient()

    policy = {
        "title": "Dynamic Zero Trust Policy",
        "rules": [
            {
                "description": "Require corp device and location",
                "device_policy": {
                    "require_screen_lock": True,
                    "allowed_device_management_levels": ["MANAGED"],
                },
                "conditions": {
                    "regions": ["US"],
                }
            }
        ]
    }

    response = client.create_access_policy(
        request={"policy": policy}
    )
    return response
```

Real-World Case Study: Using AI-Driven IAM at a Global Enterprise

A multinational Fortune 500 company, working with more than 55,000 employees, still struggles with the IAM roles. The company has manual access reviews, faces an increased number of insider threats, and has excessive permissions on their hybrid cloud environment. The company has therefore sought to use an AI-driven IAM solution, which will assist in the progressive security needs of the organization.

The company deployed predictive IAM to make automated adjustments to the access privileges based on job functions. This helped to reduce the number of privileges by an estimated 45%. Also, the organization used AI-powered aero trust to ensure the inclusion of endpoint detection tools, ensuring that credential-based breaches were cut by an estimated 70%. Finally, generative AI has helped the company to implement an automated access review, summarizing reports for auditors and ensuring compliance cycles take weeks, not months.

By using AI, the organization improved its IAM and HR processes. Onboarding and offboarding became 80% faster with AI-driven role recommendations. Security incidents from over-privileged accounts dropped by 30%. This approach also helped the company earn ISO 27001 certification by showing strong, auditable policy controls.

Generative AI and AI have exceeded expectations in companies. Using decentralized identities through AI can prevent synthetic identity fraud. Additionally, advances are needed in self-healing IAM and in Generative AI for policy management to improve AI development. These AI tools can help deliver strong results in IAM management.

References

Ait Chikh, S. (2023). FinOps: Monitoring and Controlling GCP costs.

Ajish, D. (2024). The significance of artificial intelligence in zero-trust technologies: a comprehensive review. *Journal of Electrical Systems and Information Technology*, *11*(1), 30.

Anderson, J. (2022). The Role of Identity and Access Management (IAM) in Securing Cloud Workloads.

Banu, A. (2024). AI-Powered Digital Identity Protection: Preventing Fraud in Online Transactions.

Duggal, A. K., & Dave, M. (2021). Intelligent identity and access management using neural networks. *Indian J Comput Sci Eng*.

Ike, J. E., Kessie, J. D., Okaro, H. E., Ezeife, E., & Onibokun, T. (2025). Identity and Access Management in Cloud Storage: A Comprehensive Guide.

Kumar, B., & Sharma, S. (2025). The state of cloud security: An analytical review. *Artificial Intelligence for Cyber Security and Industry 4.0*, 319-342.

Mamidi, S. R. (2024). The Role of AI and Machine Learning in Enhancing Cloud Security. *Journal of Artificial Intelligence General science (JAIGS) ISSN: 3006-4023*, 3(1), 403-417.

Olabanji, S. O., Olaniyi, O. O., Adigwe, C. S., Okunleye, O. J., & Oladoyinbo, T. O. (2024). AI for Identity and Access Management (IAM) in the cloud: Exploring the potential of artificial intelligence to improve user authentication, authorization, and access control within cloud-based systems. *Authorization, and Access Control within Cloud-Based Systems (January 25, 2024)*.

Rehan, H. (2024). AI-driven cloud security: The future of safeguarding sensitive data in the digital age. *Journal of Artificial Intelligence General science (JAIGS) ISSN: 3006-4023*, 1(1), 132-151.

Rimon, S. T., Sufian, M. A., Guria, Z. M., Morshed, N., Mosaddeque, A. I., & Ahamed, A. (2024, December). Impact of AI-powered business intelligence on smart city policy-making and data-driven governance. In *IET Conference Proceedings CP908* (Vol. 2024, No. 30, pp. 475-481). Stevenage, UK: The Institution of Engineering and Technology.

Robert, L. (2023). Integrating AI and IAM for Comprehensive Cybersecurity in GxP-Regulated Healthcare Environments.

Roy, A., Banerjee, A., & Bhardwaj, N. (2021). A study on google cloud platform (gcp) and its security. *Machine Learning Techniques and Analytics for Cloud Security*, 313-338.

Saminathan, M. (2024). *Mastering Big Data Engineering: AWS, GCP, & Azure Showdown*. Libertatem Media Private Limited.

Stutz, D., de Assis, J. T., Laghari, A. A., Khan, A. A., Andreopoulos, N., Terziev, A., ... & Grata, E. G. (2024). Enhancing security in cloud computing using artificial intelligence (AI). *Applying Artificial Intelligence in Cybersecurity Analytics and Cyber Threat Detection*, 179-220.

Sugureddy, A. R. (2024). Data governance excellence in the cloud leveraging GCP for enhanced lineage and security. *Journal ID, 2364*, 4269.

Index

A

Active Directory–based RBAC, 115
Adaptive access control, 148
AI-driven access control, 13
AI-driven IAM
    ethical considerations
        bias, 157
        compliance standards, 158
        ensure transparency and fairness, 157, 158
    predictive IAM, 158, 159
    tools in GCP, 149
    zero-trust frameworks, 159, 160
AI-driven policy management
    Cloud Audit Logs, 155
    historical access patterns, 153
    in multinational organization, 153, 154
    Policy Intelligence, 154
    policy recommendations in GCP, 152
    traditional approach, 151
AI-driven threat detection, 128
AI-powered IAM system, 148
AI-powered identity verification, 155
    biometric verification, 155
    manual identity approaches, 156
    multifactor authentication, 156, 157
AI-powered MFA system, 155, 157
Anomaly detection, 149–151
API-based SaaS product, 80
API Gateway, 69, 70, 73
    IAM roles, 70, 71
API keys, 75, 77–79
App engine, 51
Artificial intelligence (AI), 9
    anomaly detection with AI, 149–151
    in cloud security, 147, 148
    in IAM, 147
Attribute access control (ABAC), 1
Authentication, 2
Authorization, 2

B

BigQuery, 131–134, 142, 143, 153, 154
Bucket-level IAM policies, 135

C

Capital One Data Breach (2019), 28, 29
Centralization, 93, 95
Cloud Armor, 149
Cloud asset inventory (CAI), 108–111
Cloud Audit Logs, 56, 138, 150, 151, 155
    Admins Activity and data access logs, 99
    automation, 102
    centralized log management, 102
    cloud service providers, 99
    educate to teams and employees, 103
    enable process, 100
    principle of least privilege, 102
    regular review, audit logs, 102
    track IAM policy changes, 100, 101
    using multifactor authentication, 102
Cloud environments, 38

INDEX

Cloud Functions, 51, 69, 71, 76
    IAM roles, 71
Cloud Identity, 116
Cloud monitoring, 56, 104
    alert IAM strategies, 106
        automated remediation, 107
        multiproject IAM monitoring, 106
        real-time anomaly detection, 107
    IAM alerting activities
        alerting policies, 107
        employ granular filters, 107
        prioritization, high-risk actions, 107
        SIEM tools, 107
    set up IAM alerts, 105
        alerting system, 106
        alert policy, 105
        critical IAM events, 105
        log-based metrics, 105
Cloud Run, 70, 72, 76
    IAM permissions and roles, 72, 73
Cloud Spanner, 136, 137
Cloud storage, 131, 134–136, 143
CMEK encryption, 143
Company X, 80
Compute Engine, 51
Context-aware and access policies, 46
    education, 47
    healthcare industry, 46, 47
    seamless integration, 47
Cross-cloud IAM governance, 120
    access patterns and practices, 121
    automate IAM compliance checks, 122
    emergency access planning activities, 121
    implementation, centralized IdP, 120
    principle of least privilege, 120
    service accounts and workload identities, 121
    standardization, IAM policies, 120, 121
Cross-project and cross-organization access, 38–40
    benefits, 41, 42
    best practices, 42, 43
    enabling, 40, 41
Customer-managed encryption keys (CMEKs), 138–141, 143, 144
    centralized key management with KMS, 140
    disaster recovery planning, 141
    lifecycle management and key rotation interventions, 140
    monitoring, auditing and alerting, 141
    strict IAM control for keys, 140

D

Default service accounts, 51, 52

E

Encryption, 138

F

Fine-grained access control, 141–143
Fintech industry, 8, 9
Fraud prevention, 9, 11

G

Gaming industry, 10, 11
Gcloud CLI, 11, 12
Google Cloud, 69
Google Cloud Audit logs, 100
Google Cloud Console, 11, 12
Google Cloud Platform (GCP), 17, 69

custom roles, 69
predefined roles, 69
Google Managed Keys, 138
Google-managed service accounts, 55
Google service accounts, 74
Google's Vision AI, 149, 155, 157
Google Workspace, 116

H

HashiCorp Sentinel, 117, 124
Healthcare API, 143
Healthcare industry, 9, 10, 46, 47
Health Insurance Portability and
    Accountability Act (HIPAA),
    108, 111, 112

I

IAM compliance checks, 122
IAM permissions, 70
    for Cloud Functions, 71
    for Cloud Run, 72, 73
    manage service accounts, 73
    troubleshoot issues, 74
    in using API Gateway, 70, 71
IAM policies, 70–72, 74
    auditing and monitoring, 94
    automation for governance, 94
    automation, policy updates, 96
    centralization, 93, 94
    compliance and regulatory
        requirements, 95
    multitenant environments, 95
    organization policies, 94
    scripts and IAM API
        automated services, 90
        create custom role, 91
        GCP project, 89
        IAM API enabled, 89
        IAM workflows, 93
        monitoring/auditing, 92
        programming language, 89
        Python on Gcloud, 92
        RESTful API, 89
        service account JSON key, 89, 90
        update process, 90
        version control, 92
    training and awareness programs, 95
    using Cloud Audit Logs, 99
Identity and Access Management (IAM),
    1, 56, 69, 147
    auditing and monitoring access, 138
    benefits, 3, 148
    for BigQuery, Cloud Storage, and
        Spanner, 131, 132
    components, 1–3
    conditions, 33, 34
        benefits, 36
        best practices, 37
        context-aware access, 34–36
        features, 33
    configuring tools
        APIs, 12
        Gcloud CLI, 11
        Google Cloud Console, 11
        Terraform, 12
    context-aware and access policies, 46
    cross-project and cross-organization
        access, 40
    development, 4
    emerging trends
        AI-driven access control, 13
        Zero-trust principles, 13
    fintech industry, 11
        fraud prevention, 9

INDEX

Identity and Access
  Management (IAM) (*cont.*)
  regulatory compliance, 9
  secure transactions, 8
  user management, 8
  uses, 8
 gaming industry, 10, 11
 GCP, 69, 70
  identities, 5
  policies, 6, 7
  resource hierarchy, 22, 24
  resources, 6
  roles, 7
 healthcare industry, 9–11
 modern organizations, 4
 permissions, 70
 policies, 21, 22, 83
  best practices, 25, 26
  overriding, 25
  policy evaluation, 24
  policy inheritance, 23, 24
 RBAC, 17
 real-life lessons, 28–30
 roles
  basic roles, 20
  custom roles, 19
  predefined roles, 17, 18
  role management, 21
 roles in BigQuery, 132–134
 roles in Cloud Spanner, 136, 137
 roles in Cloud storage, 134–136
 Spotify, 26
 Temporary and time-bound
   access, 44
Identity-Aware Proxy (IAP), 80
Identity-centric security, 128
Infrastructure as code (IaC), 12, 83

Integration strategies
  cross-cloud service accounts, 116
  identity federation using SSO, 116
  policy-as-code approach, 117
  single sign-on process, 118
  Workload Identity Federation, 116

J, K, L

JSON Web Token (JWT), 75
Just-in-time (JIT), 44

M, N

Machine learning (ML), 9, 149
Multicloud access
  compliance on jurisdictions, 126
  identity fragmentation and siloed
    access, 123
  inconsistent permission models,
    123, 124
  lack of centralized auditing and
    compliance, 125
  manage service accounts and machine
    identities, 125
Multicloud IAM integration, 119
  conditional access policies, 119
  monitoring and auditing, 118
  principle of least privilege, 118
  user lifecycle management, 119
Multifactor authentication (MFA), 1, 8, 45

O

OAuth 2.0, 75, 76, 80
Object-level ACLs, 135
Optical Character Recognition (OCR), 156

P, Q

Policy Intelligence, 149, 152, 154
Policy Troubleshooter, 150, 151
Predictive IAM, 158, 159
Project management, 26

R

Regulatory compliance, 3, 8–10
Resource-level IAM, 142
Role-based access control (RBAC), 1, 17, 112

S

Security Assertion Markup
 Language (SAML) 2.0, 116
Security information and event
 management (SIEM) tools, 102
Service account keys, 62
 benefits, 62
 key management, 65
  automated key rotation, 66
  conditional access policies, 66
  use, 65, 66
 security risks, 62
 strategies, 62
  key rotation activities, 63
  minimal service account key use, 62
  monitoring key use, 63
  restricting access, 63
  rotation process, 63
  storage strategies, 62
Service accounts, 5, 51
 default, 51
  automatic creation, 52
  broad permissions, 51
  GCP default service accounts, 52
  limited customization, 52
 IAM permissions, 73
 strategies, 55
  auditing and monitoring, 56
  avoidance, default service
   accounts, 56
  least privilege, 55
  separate service accounts, 56
  service account impersonation, 57
  service account keys, 57
  short-lived credentials, 56, 57
  WIF, 56, 57
 types, 51
 user-managed
  code, 53
  custom roles and permissions, 53
  Google-managed, 55
  granular control, 53
  manual creation, 53
 workload identity, 57
Service-to-service authentication, 75
 API keys, 77, 78
 authorization, 75
 OAuth 2.0 *vs.* API keys, 79
 using OAuth 2.0, 75, 76
Service-to-service communication, 74, 75
SIEM tools, 107
SMS-based multifactor
 authentication, 156
Snapchat's IAM policies, 29
Software-as-a-service (SaaS) model, 80
Spanner, 131
Spotify
 conditional policies, 27
 custom roles, 27
 folder-level segmentation, 27
 multiple projects, 26
 organization-level policies, 27
 outcome, 28

167

INDEX

T

Temporary and time-bound access, 43
    demand, 44
    strategies, secure, 44, 45
    summary, 45
Terraform, 12, 83
    cross-platform support, 83
    declarative syntax, 83
    for GCP IAM Management
        auditing and monitoring, 88
        authenticate, 84
        configure GCP provider, 84
        JSON file, 84
        policy definition, 84
        resource-level management, 86
        terraform apply, 85
    IAM policies through Terraform, 88
    modularity, 83
    state management, 83
    version control, 83

U, V

Uber's GitHub IAM scandal (2016), 29
User experience, 3, 10, 11
User-managed service accounts, 52, 53

W, X, Y

Workload Identity Federation (WIF), 116, 117, 123, 125
    attribute exchange, 58
    benefits, 59
    CI/CD pipelines, 61
    enable WIF, steps and Python code, 59, 60
    external identity providers, 58
    functionality, on GCP and AWS, 58
    hybrid cloud environments, 60
    implementation, 61, 62
    Kubernetes workloads, 61
    multicloud workloads, 61
    token exchange, 58
    tools, 58
    workload identity pool, 58

Z

Zero trust, 127, 129
Zero-Trust Architecture (ZTA), 127
Zero-trust frameworks, 159
Zero-trust principles, 13

GPSR Compliance

The European Union's (EU) General Product Safety Regulation (GPSR) is a set of rules that requires consumer products to be safe and our obligations to ensure this.

If you have any concerns about our products, you can contact us on

ProductSafety@springernature.com

In case Publisher is established outside the EU, the EU authorized representative is:

Springer Nature Customer Service Center GmbH
Europaplatz 3
69115 Heidelberg, Germany

www.ingramcontent.com/pod-product-compliance
Lightning Source LLC
LaVergne TN
LVHW081450060526
838201LV00050BA/1757